Geotechnologies and the Environment

VOLUME 12

Series Editors:

Jay D. Gatrell, *College of Graduate and Professional Studies and Department of Earth and Environmental Systems, Indiana State University, Terre Haute, IN, USA*

Ryan R. Jensen, *Department of Geography, Brigham Young University, Provo, UT, USA*

The "Geotechnologies and the Environment" series is intended to provide specialists in the geotechnologies and academics who utilize these technologies, with an opportunity to share novel approaches, present interesting (sometimes counter-intuitive) case studies, and most importantly to situate GIS, remote sensing, GPS, the internet, new technologies, and methodological advances in a real world context. In doing so, the books in the series will be inherently applied and reflect the rich variety of research performed by geographers and allied professionals.

Beyond the applied nature of many of the papers and individual contributions, the series interrogates the dynamic relationship between nature and society. For this reason, many contributors focus on human-environment interactions. The series are not limited to an interpretation of the environment as nature per se. Rather, the series "places" people and social forces in context and thus explore the many sociospatial environments humans construct for themselves as they settle the landscape. Consequently, contributions will use geotechnologies to examine both urban and rural landscapes.

More information about this series at http://www.springer.com/series/8088

Amy J. Blatt

Health, Science, and Place

A New Model

Amy J. Blatt
TerraFirm International Corp., Inc.
Wayne, PA, USA

ISBN 978-3-319-12002-7 ISBN 978-3-319-12003-4 (eBook)
DOI 10.1007/978-3-319-12003-4
Springer Cham Heidelberg New York Dordrecht London

Library of Congress Control Number: 2014956216

Printed on acid-free paper

Springer is part of Springer Science+Business Media (www.springer.com)

To my husband, whose loving support made this book possible

Preface

The Patient Protection and Affordable Care Act of 2010 provides the main impetus for this book. In this volume, I attempt to re-contextualize the role of medical geographers within the history of health care reform. The PPACA presents an excellent window of opportunity for medical geographers to contribute a patient-centric view of health care that takes into account a person's place history – i.e., where a person lives, works, and plays. As such, this book presents a new model of patient care that emphasizes a patient's geographic and medical history through a broad survey of seemingly disparate disciplines, such as disease ecology, GIScience, transportation geography, cultural geography, and geographic medicine. What emerges from this volume is a deeper appreciation of the challenges of national health care reform, and the importance of a new and emerging sub-discipline of medicine, known as geospatial medicine.

While this book is, admittedly, not an authoritative tome of basic concepts and developments in medical geography, it does present enough elementary material to enable the reader to grasp the significance of the current advances in medical geography and its historical context. New and timely topics such as disease surveillance in mass gatherings and knowledge discovery with 'big data' are explored, as well as the seemingly timeless topics of environmental health, disease ecology, and quantitative GIS research.

In addition, examples of successful research programs contributing to the cutting-edge of medical geographical research are described in these pages. The Center for Geospatial Medicine and the National Children's Study are large multi-site research projects (funded by the National Institutes of Health, or NIH) that are founded on the principles discussed in this volume. In addition, the Association of American Geographers and the NIH have collaborated on a report, entitled "Establishing an NIH-Wide Geospatial Infrastructure for Medical Research: Opportunities, Challenges, and Next Steps" that describes the importance of geographic thought and leadership in health-science research.

A book on medical geography is important if it can show that the discipline provides critical, life-saving insights about patient care and public health that are vital to human health. Health care services and delivery systems are rapidly changing, as are ways in which patients are represented in these systems. As described in Chap. 2, the "ways in which patients and their medical conditions are represented, compiled and examined – from office visits and prescriptions records to daily personal blogs and newspaper stories – have also changed…" This volume goes on to explore how the diagnosis, prognosis, and treatment of chronic and infectious diseases can be significantly enhanced through the use of geospatial health intelligence in a clinical setting. The recognition of the importance of place history in clinical care has yet to be widely accepted and advocated by the medical community. Once this barrier of entry is removed, the possibilities for medical geographers to affect population health and patient care, through the use of geographic information systems (GIS) technologies, will be without bound.

I would like to acknowledge several people who provided the support and assistance necessary to the formulation, research, and publication of this volume. First, I extend deep gratitude to Frank Galgano and Sarah Wingo of Villanova University, for generously making the university's library and archival resources available to me, during the research phases of this project. In addition to the excellent support staff at Springer, I would like to call out Rajeswari Balachandran, Marielle Klijn, and Robert Doe for their superb guidance through every facet of the book publication and marketing process. I would like to thank Paige Andrew and Katherine Weimer, who recognized the amount of time and dedication needed for this project and kindly allowed me to have a more relaxed schedule during my tenure on the editorial board of their journal. Finally, I would like to thank the many friends and family members who understood the value of medical geography and its meaning in my life.

Wayne, PA, USA Amy J. Blatt, Ph.D., GISP

Contents

Part I
What Is the Role of Geography in Health Care Reform?

Chapter 1
Health Care Reform and Disease Surveillance: Making the Connection

Abstract The Patient Protection and Affordable Care Act (PPACA) was signed into law in 2010, following a history of health reforms in the United States that includes the adoption of Medicare and Medicaid in 1965 and Medicare Modernization Act in 2003. The purpose of the PPACA is to extend insurance coverage to uninsured Americans; however, it also includes provisions to improve health care quality, curb health care costs, and promote population health and wellness. The law emphasizes disease prevention through several measures, such as increased access to recommended clinical preventive services, provision of additional funding for prevention programs, and the coordination of patient care through the use of health information exchanges (HIEs). Through a stronger collaboration between public health and health care systems, the use of HIEs presents a unique opportunity for professional geographers to collaborate with other scientists on disease surveillance systems that allow an increased focus on disease prevention and health outcomes.

Keywords Health care reform • Disease surveillance • Patient Protection and Affordable Care Act • Health information exchange • Population health • Disease management • Medical geography

1.1 Introduction

When President Barack Obama signed the Patient Protection and Affordable Care Act (PPACA) into law on March 23, 2010, the die was cast to invest in the nation's health programs on chronic disease prevention, health promotion, and public health [1]. Through Title IV of the PPACA, $2 billion in annual funding was provided to the National Prevention, Health Promotion, and Public Health Council. In addition, the PPACA directed the Secretary of Health and Human Services (HHS) to support research innovations in disease prevention, intervention, and community-based public health programs [2]. These financial investments renewed the support of health research on rural and under-represented minority groups, health care disparities in the Medicaid and Children's Health Insurance Programs (CHIP) populations, and employer-based wellness public health surveillance programs.

In addition, the PPACA specified a "Strengthening Public Health Surveillance Systems" program that established an Epidemiology and Laboratory Capacity

A.J. Blatt, *Health, Science, and Place*, Geotechnologies and the Environment 12,
DOI 10.1007/978-3-319-12003-4_1

Grant Program to "assist public health agencies in improving surveillance for, and response to, infectious diseases" [3]. The new law also directed the Office of the National Coordinator of Health Information Technology (ONC-HIT) to "develop national standards for the management of data collected," and "develop interoperability and security systems for data management" – i.e., to implement and operationalize a health information exchange (HIE) [4].

These health care reform initiatives – along with the 2009 American Recovery and Reinvestment Act (ARRA) which expanded the national broadband program to deliver high-speed telecommunications access to unserved and underserved areas – allow health care providers, organizations, and insurers to mine and interpret health care data for innovative ways to improve patient care and outcomes at a reduced cost [5]. As public and private health information exchange organizations (HIOs) are becoming more integrated into the national infrastructure and health information networks, there arises numerous opportunities for professional geographers to advance clinically relevant research in the areas of health care services, delivery, and access.

1.2 A Brief History of Health Care Reform in the United States

Since the early twentieth century, no fewer than ten presidents have attempted to enact legislation to reform health care and insurance coverage in the United States. They are: Presidents Franklin Roosevelt, Harry Truman, John Kennedy, Lyndon Johnson, Richard Nixon, Jimmy Carter, Ronald Reagan, Bill Clinton, George W. Bush, and Barack Obama (see Table 1.1 for a list of major health care legislations in United States history). In 1912, presidential candidate Theodore Roosevelt presented the issue of national health coverage to the general public, as he unsuccessfully ran for president as a candidate of the Bull Moose Party. His distant cousin, President Franklin Roosevelt, sought to include health insurance coverage in the Social Security Act of 1935. However, strong opposition from the American Medical Association (AMA) to institute socialized medicine against a backdrop of the Great Depression forced him to withdraw it. In 1943, he crafted a new piece of national health insurance legislation but passed away in 1945, before the bill could be introduced [6].

Table 1.1 Landmark health care legislations, by year and presidency

President	Year	Legislation or program
Franklin Roosevelt	1935	Social Security Act
Lyndon Johnson	1965	Medicare and Medicaid
Ronald Reagan	1985	Consolidated Omnibus Budget Reconciliation Act (COBRA)
William Clinton	1997	State Children's Health Insurance Program (SCHIP)
George W. Bush	2003	Medicare Modernization Act (MMA)
Barack Obama	2009	Health Information Technology for Economic and Clinical Health Act (HITECH)
Barack Obama	2010	Patient Protection and Affordable Care Act (PPACA)

As President Roosevelt's successor, President Harry Truman championed a national health insurance plan that was similar to a Canadian-style single payer public health insurance system (i.e., a national insurance program covers all of those who pay voluntary fees). He was not successful at this. Towards the end of his tenure as president, President Truman began to work on another piece of legislation to provide health insurance coverage for senior citizens. President John Kennedy continued to carry the torch for a health insurance for the nation's senior citizens, but was assassinated in 1963, before this work could be completed.

In 1965, a national Medicare and Medicaid program became law in the United States, during President Lyndon Johnson's administration [6]. Medicare was a federal health insurance program for three populations: (1) those who are 65 or older, (2) certain younger people with disabilities, and (3) people with permanent kidney failure requiring dialysis or a transplant, or end-stage renal disease (ESRD). When it was first created, Medicare contained two parts. Medicare Part A covered the costs associated with hospital care, skilled nursing facility care, hospice care, and home health care. Medicare Part B covered the costs associated with health care providers, outpatient care, home health care, and durable medical equipment. To address people with limited income and resources, Medicaid was created to help with medical costs for this population, and covered services not normally covered by Medicare, such as long-term supports and services, and personal care services (Fig. 1.1).

After President Lyndon Johnson left the White House, President Richard Nixon embraced the goal of universal health insurance and proposed a plan that: (1) provided private coverage for most Americans (including subsidies); (2) mandated employers provide insurance to their workers; and (3) required all individuals to enroll. Unfortunately, these efforts took a back stage to the Watergate scandals, and

Fig. 1.1 Centers for Medicare and Medicaid Services website (http://www.cms.gov/)

President Nixon's plan was never legislated. In June 1979, President Jimmy Carter sought to reform the medical system by phasing in a system to lower physician and hospital costs, but the bill was defeated in the House, halfway into the third year of his presidential tenure [6].

As President Carter's successor, President Ronald Reagan enacted the next large piece of health care legislation in 1985, by signing The Consolidated Omnibus Budget Reconciliation Act of 1985 (COBRA) into law (Fig. 1.2). COBRA required all employers to cover former employees on the company health plan for 18 months after leaving a job, with the employees bearing the cost of this coverage [7].

Later, in 1993, President Clinton and First Lady Hillary Rodham Clinton proposed a health care reform package that mandated universal coverage, health insurance reform, regional alliances among health insurance plans, and consumer choice of health plans. It was called the Health Security Act of 1993. A core element of this plan was to mandate that employers provide health insurance coverage to all employees through competitive, but closely regulated, health maintenance organizations [8]. Because it faced strong political opposition, the final Democratic bill did not become law.

In 1997, however, President Clinton did sign a bill into law that created the State Children's Health Insurance Program (SCHIP). SCHIP was the largest expansion of taxpayer-funded health insurance coverage for American children since the inception of Medicaid. Administered by the United States Department of Health and Human Services (DHHS), the program provided matching funds to states for health insurance to families with children, especially those in families with incomes that are modest but too high to qualify for Medicaid [9].

Since the creation of the Medicare and Medicaid programs in 1965, the role of prescription drugs in health care has increased significantly. The introduction of

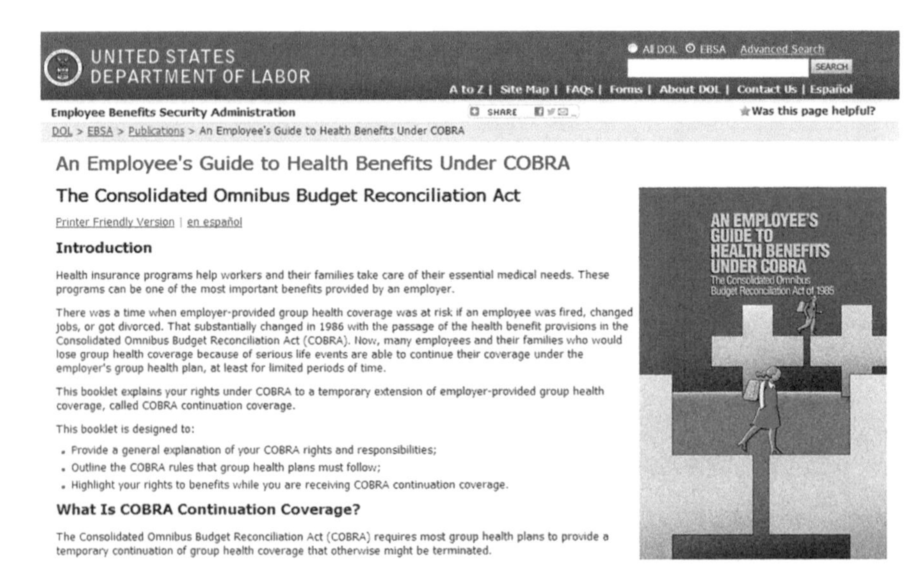

Fig. 1.2 An employee's guide to health benefits under COBRA, on the U.S. Department of Labor website (http://www.dol.gov/ebsa/publications/cobraemployee.html)

new and expensive drugs has made prescriptions less affordable for many patients, particularly senior citizens. To address this growing problem, President George W. Bush signed The Medicare Prescription Drug, Improvement, and Modernization Act, or the Medicare Modernization Act (MMA) into law, in 2003. The MMA created a prescription drug benefit (called Medicare Part D) in 2006, and required that insurance companies and HMOs providing this voluntary benefit support electronic subscribing. The MMA also created new Medicare Advantage – or Medicare Part C – plans to allow private insurers to provide Part A, B, and D benefits (i.e., benefits that were traditionally provided by the government). Additionally, the MMA introduced numerous other programs, such as the Health Savings Account statute that expanded allowable contributions and employer participation [10].

Despite these ambitious programs, the number of uninsured and underinsured Americans has increased every year. In the 1990s, approximately 37 million Americans lacked insurance coverage. By 2010, the number had risen to 50 million [11]. Also, 25 million adult Americans were underinsured in 2007, up 60 % from 2003. The underinsured have limited health insurance coverage with cost-sharing requirements that place them at financial risk, in the event of a serious illness or injury. Many Americans – as many as 45,000 people – die each year because they do not have adequate health insurance [12, 13]. In addition, illness and medical bills contribute to a large and increasing share of bankruptcies in the United States [14].

To address these health care problems, President Barack Obama signed the PPACA into law on March 23, 2010. The legislation's key components required that: (1) almost everyone have health insurance coverage or pay a fine; (2) most companies provide health insurance coverage to their workers; (3) insurance companies accept all applicants, regardless of any pre-existing conditions; and (4) subsidies be provided to those who can't afford insurance. On June 28, 2012, the Supreme Court upheld the constitutionality of the PPACA, in a 5–4 decision, and its much-disputed mandate that almost all Americans have health coverage or pay a fine [15].

1.3 Role of HIEs in Population Health and Disease Management

In the wake of President Obama's re-election in 2012 and the favorable Supreme Court ruling, there was little doubt in people's minds that the PPACA would drive the implementation of HIEs (see Table 1.2 for an explanation of commonly used acronyms in this book). In extending insurance coverage for 32 million uninsured people, the PPACA expanded the number of care settings available in the health care system and generated a greater demand for the HIEs. It is anticipated that these HIEs will allow providers to coordinate care and access patients' records in multiple different settings. The Health Information Technology for Economic and Clinical Health (HITECH) Act supports the demand for the HIEs through several incentives and grant programs. Health care providers are eligible for up to $44,000 for using electronic health records (EHRs) in a meaningful way by 2015; otherwise, they would incur financial penalties [16].

Table 1.2 Health care acronyms frequently used in this book

Acronym	Explanation
AMA	American Medical Association
ARRA	American Recovery and Reinvestment Act
CHIP	Children's Health Insurance Programs
COBRA	Consolidated Omnibus Budget Reconciliation Act
EHRs	Electronic health records
ESRD	End-stage renal disease
GIS	Geographic information systems
HHS	Health and Human Services
HIEs	Health information exchanges
HIOs	Health information exchange organizations
HITECH	Health Information Technology for Economic and Clinical Health
IOM	Institute of Medicine
MMA	Medicare Modernization Act
NHANES III	Third National Health and Nutrition Examination Survey
ONC-HIT	Office of the National Coordinator of Health Information Technology
PPACA	Patient Protection and Affordable Care Act
SCHIP	State Children's Health Insurance Program

Supporters of the PPACA believe that the HIEs will advance five health care goals: (1) improving the quality, safety, and efficiency of care while reducing disparities; (2) engaging patients and families in their care; (3) promoting public and population health; (4) improving care coordination; and (5) promoting the privacy and security of EHRs (Fig. 1.3).

This growing attention on improving personal and population health outcomes through the HIEs provides the context and motivation for this book on a new model for geography in medicine. The purpose of this book is to re-cast the role of medical geography, especially as it pertains to health care services, access, and delivery. Through the study and application of volunteered geographic information, geospatial data integration, and the interplay between geographic and geospatial medicine, the first part of this book presents geographers in the unique role of addressing the key scientific and public health questions posed by the PPACA. The second part of this book explores how volunteered geographic information, social media networks, and public participation can contribute to our understanding of disease surveillance and management. By applying the lens of the landscape as an organizing principle, ways to represent health and disease in patients will be examined. In addition, this section will also discuss how geographic knowledge and thought helps us to understand pluralistic and minority populations, as well as the importance of public participation and its role in disease epidemiology in mass gatherings.

The third part of this book focuses on the technological opportunities and challenges – arising from the HIEs – that need to be addressed, in order for true geospatial data integration to exist within our current health care delivery system. The proliferation of an application-driven internet delivery system, cloud technology, and mobile smart devices contribute to the exponential growth in data available for

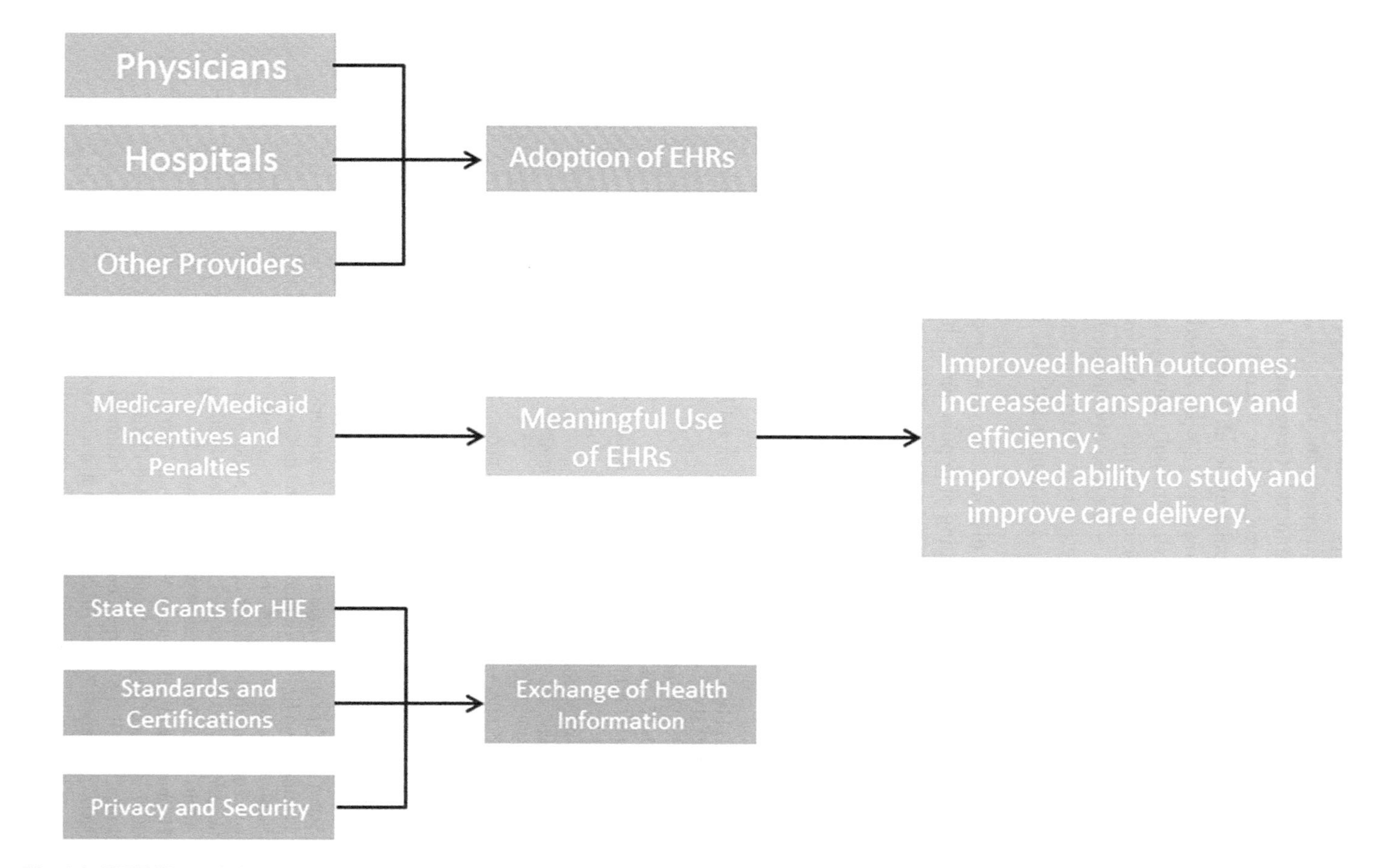

Fig. 1.3 HITECH Act's framework for meaningful use of EHRs (Adapted from: Blumenthal [17])

geospatial health analyses. This section will address how this flood of data from different sources affects our ability to track a patient's personal and medical histories across different jurisdictions (i.e., geomedicine).

The fourth part of this book, entitled "Geography in Medicine," begins with how theories of human movement can inform scientists and epidemiologists about disease transmission at the population level. It ends with a discussion of a new sub-discipline of medicine, known as geospatial medicine, which examines the genetic, social, cultural, and environmental pathways of disease pathogenesis and transmission. Several major research centers – such as those at the University of Michigan and Duke University – are heavily invested in this important research initiative. The purpose of this section is to underscore the importance of the vast and multidimensional complexities of human health and development, especially in context of disease management.

Before embarking on this discourse in how geographic thought plays an influential role in our understanding of population health and disease management, it is important to first recognize the origins of disease surveillance – and its rich history in disease prevention efforts and the American public health arena.

References

1. H.R. 3590–111th Congress: Patient Protection and Affordable Care Act. (2009)
2. PPACA Title IIV
3. PPACA Title IV, Subtitle C
4. PPACA Title XXXI, Section 3101
5. Rosenbaum S (2011) Law and the public's health. Public Health Rep 126:130–135
6. Blumenthal D, Morone J (2009) The heart of power: health and politics in the Oval Office. University of California Press, Berkeley
7. Ross M, Hayes C (1986) Consolidated Omnibus Budget Reconciliation Act of 1985. Soc Secur Bull 49(8):22–31
8. Mariner WK (1994) Patients' rights to care under Clinton's Health Security Act: the structure of reform. Am J Public Health 84(8):1330–1335
9. Committee on Child Health Financing (2007) State Children's Health Insurance Program achievements, challenges, and policy recommendations. Pediatrics 119:1224–1228
10. Bell DS, Friedman MA (2005) E-Prescribing and the Medicare Modernization Act of 2003. Health Aff 24(5):1159–1169
11. U.S. Census Bureau; Current Population Survey, "Table H101: Health Insurance Coverage Status and Type of coverage by Selected Characteristics: 2009"
12. Committee on Health Insurance Status and Its Consequences, Board on Health Care Services, Institute of Medicine of the National Academies (2009) America's uninsured crisis: consequences for health and health care. National Academies Press, Washington, DC
13. Wilper AP, Woolhandler S, Lasser KE et al (2009) Health insurance and mortality in US Adults. Am J Public Health 99(12):2289–2295
14. Himmelstein BU, Thorne D, Warren E et al (2009) Medical bankruptcy in the United States, 2007: results of a national study. Am J Med 122(8):741–746
15. Liptak A (2012) Supreme Court Upholds Health Care Law, 5–4, in Victory for Obama. The New York Times, 29 June 2012. Available at: http://www.nytimes.com/2012/06/29/us/supreme-court-lets-health-law-largely-stand.html?pagewanted=all&_r=0. Accessed 31 Aug 2013
16. Blumenthal D (2011) Implementation of the federal health information technology initiative: part two of two. N Engl J Med 365:2426–2431
17. Blumenthal D (2010) Implementing HITECH. NEJM 362(5):384

Chapter 2
The Importance of Geography in Disease Surveillance

Abstract The current attention on health care reform in the United States provides an excellent opportunity for professional medical geographers to be engaged in research on population and community health. As the history of disease surveillance in the United States indicates, there is a need for more synthetic and integrative research on disease surveillance systems that can improve health outcomes and quality of care. Such a system would incorporate the principles of an accessible and distributed surveillance infrastructure and multiple streams of data based on shared references to the common geographic locations. Medical geographers are well poised to address the technical demands of these issues, through their knowledge of issues such as spatial data quality and resolution, the legal and ethical complexities of volunteered geographic health information, and the proliferation of web technologies (like Web 2.0 and 3.0). The research methods needed to address these topics span a number of paradigms, from the technical dimensions of GIScience to the social critiques of contemporary human geography – and have the power to engage a broad cross-section of professional geographers.

Keywords Infectious disease • Disease registries • Population health • Public health disease surveillance • Spatial data resolution • Volunteered geographic information

2.1 Introduction

One of the compelling forces behind this book is the current attention on health care reform in the United States. The health information exchanges (HIEs) created by PPACA are designed to improve the delivery of health care services by expediting the delivery of health information and increasing health care access [1–4]. Disease surveillance is an application of the HIEs that is designed to prevent, monitor, and respond to disease outbreaks.

At the most basic level, disease surveillance can be divided into three parts: detecting, understanding, and responding to the spread of disease [5]. A sophisticated disease surveillance system manages "health-related data and information for early warning of threats and hazards, early detection of events, and rapid

A.J. Blatt, *Health, Science, and Place*, Geotechnologies and the Environment 12,
DOI 10.1007/978-3-319-12003-4_2

Table 2.1 Opportunities and challenges for professional geographers in disease surveillance

Disease diffusion in areas that are not well characterized
Transport of disease vectors between certain population centers
Surveillance capacity building in resource-limited countries
Forecasting future health hazards
Geospatial data infrastructures in health information exchanges
Volunteered geographic information in public health applications
Data sharing, privacy, and ethical issues
Representation of "minority" patients and their "hidden" medical conditions
Synthetic and distributed disease surveillance systems
Geospatial data integration from multiple sources

characterization of the event so that effective actions can be taken to mitigate adverse health effects" [6].

Despite the large body of published research in medical geography and spatial epidemiology [7–10], and public health biosurveillance [11–13], there is relatively little published research addressing the design and implementation of a pro-active, spatio-temporal, disease surveillance system that incorporates the health and environmental factors necessary for a rapid response to and recovery of a disease outbreak. This is because of the difficulties in designing a system that successfully integrates and coordinates many moving parts across multiple scales and many different governance bodies.

The goal of this chapter is to demonstrate that the skills and experiences of professional geographers are essential to the field of disease surveillance. To accomplish this, it will briefly describe the history of disease surveillance in the United States. Then, it will present a survey of geospatial applications in disease surveillance, and highlight the current opportunities and challenges in this field for the professional geographers (Table 2.1).

2.2 A Brief History of Public Health Disease Surveillance in the United States

The earliest records of disease surveillance in the United States are from the colonial era, or eighteenth century, when reports of disease and infections were made to local officials by family members, lodging proprietors, and ship owners and operators. These citizens reported acute infectious disease – like cholera, yellow fever, and smallpox – so that their communities would know about possible epidemics and undertake the necessary control measures [14].

As the use of vital statistics and registries of birth, death and marriage became more prominent, disease reporting became an important way to prevent certain infectious diseases. In the late nineteenth century, Henry Bowditch and members of the state board of health in Massachusetts asked physicians to report on all infectious diseases detected, on a weekly basis [15]. As the discoveries of Louis Pasteur and Robert Koch enhanced the public's understanding of the role that bacteriology played in disease, it provided a greater role for laboratory diagnostics in disease reporting. Consequently, disease reporting shifted from the domain of sanitarians and social reformers to physicians and diagnostics laboratories. Henceforth, the bacteriological revolution played a vital role in the creation of public health departments and the importance of disease surveillance [16].

To many, disease surveillance offered the hope of vital health care services, lifesaving knowledge, and protection for the individual and community, in the face of epidemics. To others, disease surveillance was regarded as an intrusion of personal privacy and the doctor-patient confidentiality. As public health disease surveillance began to unfold in the late nineteenth century (with the surveillance of tuberculosis and venereal diseases), the American Medical Association (AMA) code of ethics evolved to consider community health prevention, as well as patient privacy rights. In 1903, the AMA code of ethics explicitly permitted exceptions to the doctor-patient confidentiality, especially when "imperatively required by the laws of the state" [17]. In 1912, the AMA code of ethics specifically stated that "A physician may not reveal the confidentiality entrusted to him… unless he is required to do so by law or unless it becomes necessary in order to protect the welfare of the individual or the community" [18].

The debates concerning public health disease surveillance took place against a backdrop of wartime and postwar repression. World War I inspired and reinvigorated discussions on civil liberties and personal freedoms. The Civil Liberties Bureau was created in 1917, and was renamed the American Civil Liberties Union in 1920 [19]. In 1959, the United States Supreme Court upheld the right of the Maryland health department to arrest and fine a homeowner who refused a search for rat infestation because the health inspector did not have a warrant [20].

Furthermore, Cold War concerns provided the catalyst to expand disease surveillance efforts at the federal level. In 1951, the threats of bioterrorism during the Korean War prompted Alexander Langmuir to establish the Epidemic Intelligence Service (EIS) at the Communicable Disease Center (later renamed to be the Centers for Disease Control and Prevention, or CDC) [21]. The EIS facilitated greater interactions between federal and state health departments, and allowed the CDC to play a more prominent role in the coordination and funding of state public health surveillance.

In 1977, the United States Supreme Court ruled in its first public health surveillance case and stated that disease reporting is "an essential part of modern medical practice," citing venereal disease, child abuse, and fetal death reports as legitimate examples of public health reporting [22]. However, with the identification of Human Immunodeficiency Virus/Acquired Immunodeficiency Syndrome (HIV/AIDS) in the 1980s, the tensions between patients' privacy and public health reporting began to re-escalate. It wasn't until 1994, when the antiretroviral agent AZT was shown to reduce HIV transmission between parent and child that the public sentiment against HIV case surveillance began to wane [23]. In September 1997, the CDC formally called upon all states to adopted a system of HIV case reporting; the move was backed by the AMA and the editorial staff of the New England Journal of Medicine [24, 25]. To allay concerns from the ACLU and patients advocacy groups, the 1999 CDC guidelines stated that any receipt of federal funds for HIV/AIDS surveillance would be contingent upon a demonstration of acceptable security standards to protect patients' identities (e.g., by using uniquely coded identifiers) [26].

Concurrently, in the 1990s, the CDC created the National Environmental Public Health Tracking Program, aimed at building a nationwide network for studying human exposures to environmental hazards (Fig. 2.1). In 2002, Congress allocated funds to help create a system that linked hazard exposure and disease data, and by 2004, the CDC announced funding for 21 local health departments and public health schools to help move toward a national environmental public health tracking system [27]. These attempts to integrate the surveillance of environmental hazards mirrored the progress evident in establishing immunization and birth defects registries. In 1999, the Department of Health and Human Services (DHHS) began funding a program to foster linkages between newborn screenings for genetic defects and other maternal and child health services. By 2003, 12 states and large metropolitan

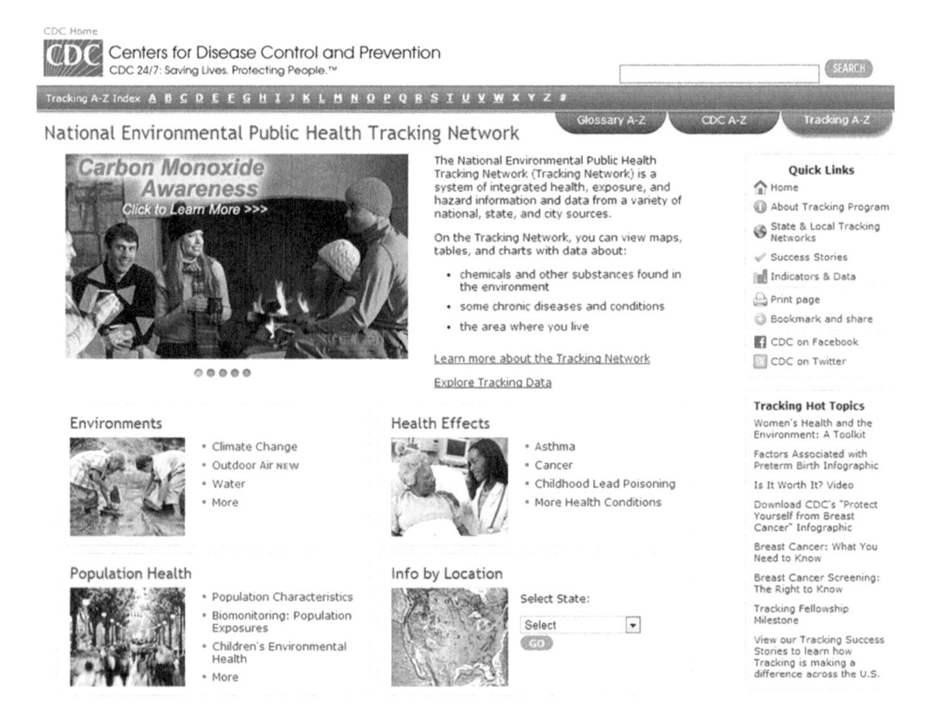

Fig. 2.1 CDC's National Environmental Public Health Tracking Network website (http://ephtracking.cdc.gov/showHome.action)

areas were establishing integrated child health data systems. Most were focused on linking data on four procedures: vital registration, newborn dried blood spot screening, immunization, and early hearing detection and intervention [28, 29]. Among the most advanced of these systems was Rhode Island's KidsNet, which combined data from nine public health programs to create a child health profile [30].

Looking back on the twentieth century, financial support for the myriad of laws and regulations mandating reporting for a wide range of illnesses had been lacking. Despite the fact that states bore the legal responsibility for determining which diseases should be reportable, they contributed only about a quarter of the resources for surveillance activities [31]. The impact of funding on the completeness rates of reporting of infectious diseases was also evident. The average rate for the high-profile and well-funded surveillance systems covering AIDS, tuberculosis, and sexually transmitted diseases was significantly better than for all other diseases (79 % versus less than 50 %) [32]. Therefore, the financial investments in chronic disease prevention and disease surveillance specified in the Patient Protection and Affordable Care Act (PPACA) of 2010 represent an opportune moment for professional geographers to contribute to the scientific understanding and response to disease outbreaks.

2.3 Geospatial Applications in Disease Surveillance: Current Opportunities and Challenges

After September 11, 2001, the nation's inadequate surveillance capabilities to counter the threat of bioterrorism became apparent. In a study conducted in early 2002, the General Accounting Office concluded that "existing surveillance systems have weaknesses, such as chronic underreporting and outdated laboratory facilities, which raise concerns about the ability of the state and local agencies to detect emerging disease or a bioterrorism threat" [33]. In October 2002, the CDC announced plans for a national syndromic surveillance system as part of the nation's defense against terrorism. Several cities, like New York City and Pittsburgh, also began to develop Early Aberration Reporting Systems (EARS) which were designed to monitor unusual clusters of illness that might suggest a bioterrorist attack [34].

As stated earlier, disease surveillance can be divided into three major components: detecting, understanding, and responding to the spread of disease. Syndromic surveillance systems are designed to detect a disease before an actual diagnosis is made, by using ancillary data sources to predict a clinical outbreak. They act like smoke detectors, sounding alarms before a disease sweeps through a community. The data sources present information from different points along the continuum of the disease process, and allow public health officials to provide health care facilities and professionals with advanced notice of a disease outbreak. Examples of these data sources include sales records of prescriptions and other disease-related consumer products (such as Kleenex tissues, Tropicana orange juice and over-the-counter medicines), work and school absenteeism records, and increased visits and calls to health care facilities and providers [35].

The second aspect of disease surveillance is understanding the spread of disease. Given the proliferation of geographic information systems (GIS) technology and advanced visualization algorithms, medical geographers have contributed much to this field [8–10, 36]. In addition, geospatial scientists have grappled with issues such as the appropriate data resolution, spatial epidemiological modeling, and developing forecasting models [5, 37–44]. As the spread of disease varies across different geographical landscapes, it is often difficult to understand the spatio-temporal diffusion of disease in areas that are not well characterized geographically. Not surprisingly, sentinel clinics have long been used to monitor certain diseases in resource-limited areas [45, 46]. The patterns of travelers and their visits to sentinel clinics can provide insights into regional illness patterns [47–51]. However, there is much uncertainty surrounding the transport of disease vectors between certain population centers [52–55].

The prevention of future disease outbreaks is the third aspect of disease surveillance. Due to challenges in information technology, infrastructure, public health resources, and the costs of proprietary software, resource-limited countries are not as able to respond to disease outbreaks as developed countries. To address these needs, the Suite for Automated Global Electronic bioSurveillance (SAGES) was developed (Fig. 2.2). SAGES is a collection of modular, flexible, freely-available software tools for electronic disease surveillance. One or more SAGES modules may be used in concert with existing surveillance applications or the entire SAGES suite may be used en masse for an end-to-end biosurveillance capability [56].

Other systems – such as BioSense [57], Distribute [58], EARS [59], and Real-time Outbreak and Disease Surveillance (RODS) [60] – collect, analyze, and display surveillance data so that different health regions and jurisdictions can compare the progression of infectious disease outbreaks. A critical element of disease surveillance systems is the ability to prevent further outbreaks by forecasting the occurrence of a health hazard. BioSINE accomplishes this by providing geospatial and temporal visualization capabilities to traditional surveillance systems. BioSINE is a project funded by the Army's Telemedicine and Advanced Technology Research Center (TATRC) to rapidly explore, analyze and share trends in public health data. Researchers can use this system to form hypothesis based on the patterns discovered to initiate additional studies that predict future outbreaks [61].

As stated at the beginning of this chapter, the goals around the national HIEs are to expedite the delivery of health information and increase health care access. In the United States, approximately 25 % of people with HIV infection don't know their HIV status. Moreover, of those who are aware that they're infected, 50 % are not receiving regular HIV care [62].

In March 2011 the CDC convened a "Consultation on Monitoring the Use of Laboratory Data Reported to HIV Surveillance" meeting to develop recommendations for the legitimate uses of confidential surveillance data in improving patient outcomes. In order to open up the surveillance registries – potentially giving infected

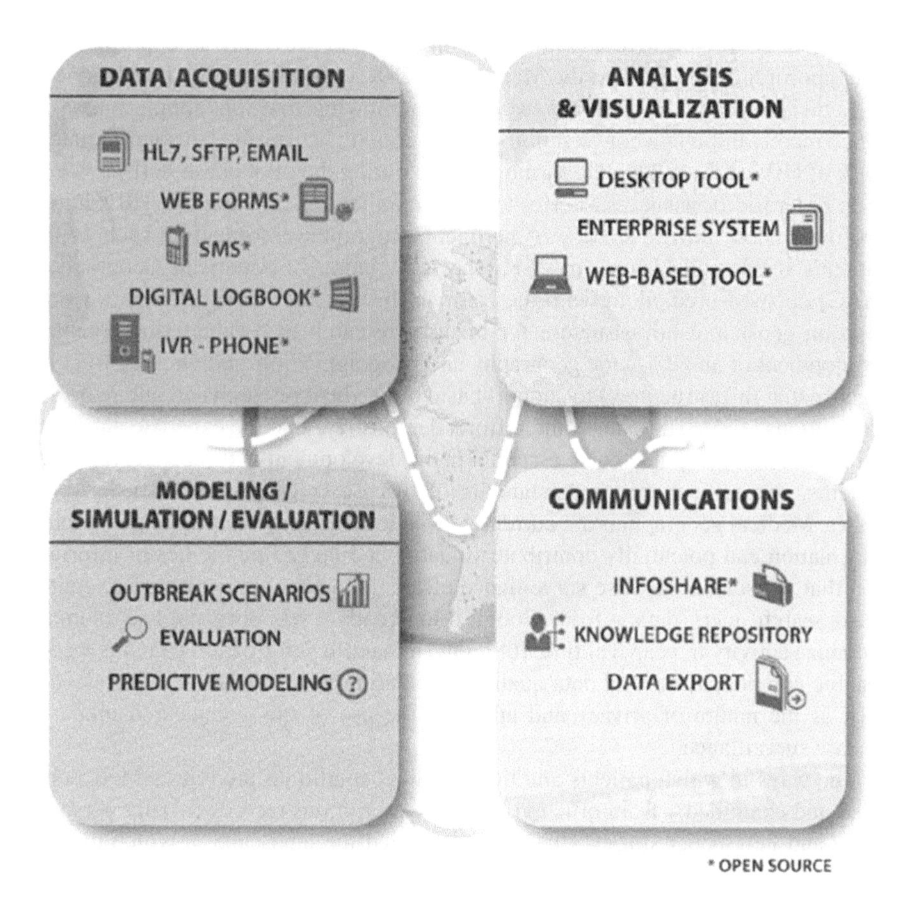

Fig. 2.2 Suite for Automated Global Electronic bioSurveillance (SAGES) – a collection of modular, flexible, and freely available software tools for electronic disease surveillance in resource-limited settings. One or more tools may be used in concert with existing surveillance applications or be used en masse for an end–to-end biosurveillance capability (From Figure 1 in: Lewis et al. [58])

people and clinicians more treatment options – the "Jericho-like" walls used to protect patient confidential must be dismantled [63].

Recognizing the controversies involved in disclosing information from surveillance registries, the state of Louisiana consulted with members of the community, health care providers, and federal health officials on these related ethical matters. As a result, the state Office of Public Health (OPH) created and implemented the Louisiana Public Health Information Exchange (LaPHIE), which allows an "authorized medical provider" to open a patient's electronic medical record in the state hospital system. The LaPHIE determines whether the patient is an HIV-exposed infant or someone who tested positive for HIV but was either not informed of the results or hasn't received a CD4 test within the past 12 months. In these instances, the system returns a "point-of-care message," alerting the caregiver that the patient is HIV-positive and not receiving care and providing an opportunity to offer appropriate services. At the March 2011 CDC consultation, Dr. Jane Herwehe of Louisiana State University presented data showing that this simple message, which merely initiates a conversation with the patient, has resulted in approximately 75 % of HIV-positive people returning to care during the pilot phase [64].

In order for these success stories to occur in other critical areas, the HIEs initiative must allow for the storage of pertinent geographic information, such as the patient's residential history. In 2011, the Association of American Geographers (AAG) co-sponsored, along with the National Institutes of Health (NIH), a workshop on geospatial infrastructure for medical research to "evaluate the potential development of an NIH-wide geography and geographic information infrastructure ("geospatial infrastructure") to support basic biomedical research and public health applications" [65]. These and other efforts demonstrate that the expertise and tools of professional geographers are essential in the development of disease surveillance systems, primarily in our understanding of disease detection and diffusion processes. Medical geographers are currently investigating how volunteered geographic information can potentially contribute valuable geo-tagged public health information that is useful to disease surveillance efforts [11, 66, 67]. For instance, aggregated search query data – from Google Flu Trends – has been used to estimate influenza activity in near-real time [68]. Given that this information contains a geographic component, several data quality and sharing issues need to be addressed – such as the nature of privacy and ethics in the use of this volunteered mode of disease surveillance.

The ways in which patients and their medical conditions are represented, compiled and examined – from office visits and prescriptions records to daily personal blogs and newspaper stories – have also changed the landscape in which disease surveillance systems operate. While volunteered geographic information will more likely contain the contextual milieu in which a medical condition is noted, the question of whether minority populations or "hidden" conditions are ignored remains unresolved.

In addition, the proliferation of web technologies, such as Web 2.0/Web 3.0 and mashups, holds promising frontiers for medical geographers to develop a new research agenda for disease surveillance that incorporates the principles of an accessible and distributed surveillance infrastructure, integrating multiple sources of data based on shared references to the common geographic locations.

In conclusion, this chapter has demonstrated a need for further integrative research on disease surveillance by the community of professional medical geographers. Topics such as spatial data quality and resolution, the legal and ethical issues of volunteered geographical health information, and the technical demands of formulating a synthetic and integrative disease surveillance system represent the types of research that only geographers can undertake. The research methods needed to address them span a number of paradigms, from the technical dimensions of GIScience to the social critiques of contemporary human geography. Consequently, these questions have the power to engage a broad cross-section of professional geographers.

References

1. Friedman DJ, Parrish RG II (2011) The population health record: concepts, definition, design, and implementation. J Am Med Inform Assoc 17:359–366
2. Roski J, McClellan M (2011) Measuring health care performance now, not tomorrow: essential steps to support effective health reform. Health Aff 30(4):682–689
3. Cebul RD, Love TE, Jain AK, Hebert CJ (2011) Electronic health records and quality of diabetes care. N Engl J Med 365:825–833
4. Romano MJ, Stafford RS (2011) Electronic health records and clinical decision support systems: impact on national ambulatory care quality. Arch Intern Med 171:897–903
5. Jiang X, Cooper GF (2007) A Bayesian spatio-temporal method for disease outbreak detection. J Am Med Inform Assoc 17:462–471
6. Centers for Disease Control and Prevention (2010) National biosurveillance strategy for human health – version 2.0. Department of Health and Human Services, Washington, DC
7. Boulus MNK (2004) Towards evidence-based, GIS-driven national spatial health information infrastructure and surveillance services in the United Kingdom. Int J Health Geogr 3:1
8. Cheung K-H, Yip KY, Townsend JP et al (2008) HCLS 2.0/3.0: Health care and life sciences data mashup using Web 2.0/3.0. J Biomed Inform 41(5):694–705
9. Cromley EK, McLaffery S (2012) GIS and public health. Guilford Press, New York
10. Meade M (2010) Medical geography. Guilford Press, New York
11. Brownstein JS, Freifeld CC, Madoff LC (2009) Digital disease detection – harnessing the web for public health surveillance. N Engl J Med 360(21):2153–2157
12. Horst MA, Coco AS (2010) Observing the spread of common illnesses through a community: using Geographic Information Systems (GIS) for surveillance. JABFM 23(1):32–41
13. Lombardo J, Burkom H, Elbert E et al (2003) A system overview of the electronic surveillance system for the early notification of community-based epidemics (ESSENCE II). J Urban Health Bull NY Acad Med 80:i32–i42
14. Thompson RH (2010) Laws relative to quarantine and to the public health of the City and port of New-York. Nabu Press, Charleston
15. Trask JW (1914) Vital statistics: a discussion of what they are and their uses in public health administration, Public health reports. Supplement no. 12. Government Printing Office, Washington, DC
16. Boyce LL (1902) The health officer's manual and public health law of the state of New York. Albany, New York
17. Gelman R (1984) Prescribing privacy: the uncertain role of the physician in the protection of privacy. North Carolina Law Rev 62:255
18. Mooney G (1999) Public health versus private practice: the contested development of compulsory notification of infectious disease in late-nineteenth century Britain. Bull Hist Med 73:238–267
19. Gormley K (1992) One hundred years of privacy. Wisconsin Law Rev 1335–1441
20. Douglas W (1962) The right of the people. Pyramid Books, Salem
21. Langmuir AD, Andrews JM (1952) Biological warfare defense. Am J Public Health Nations Health 42(3):235–238
22. Whalen v. Roe, 429 U. S. 589; 97 S. Ct. 869; 51 L. Ed. 2d. 64
23. Connor EM, Sperling RS, Gelber R et al (1994) Reduction of maternal-infant transmission of human immunodeficiency virus type 1 with zidovudine treatment. N Engl J Med 331:1173–1180
24. Centers for Disease Control and Prevention (1997) Update: trends in AIDS incidence – United States, 1996. Morb Mortal Wkly Rep 46(37):861–867
25. Steinbrook R (1997) Battling HIV on many fronts. N Engl J Med 337:779–781
26. Fleming PL, Ward JW, Janssen RS et al (1999) Guidelines for national Human Immunodeficiency Virus case surveillance, including monitoring for Human Immunodeficiency Virus infection and Acquired Immunodeficiency Syndrome. Morb Mortal Wkly Rep 48(RR13):1–28

27. McGeehin MA, Qualters JR, Niskar AS (2004) National environmental public health tracking program: bridging the information gap. Environ Health Perspect 112(14):1409–1413
28. Linzer DS, Lloyd-Puryear MA, Mann M et al (2004) Evolution of a child health profile initiative. J Public Health Manag Pract 10:S16–S23
29. Fehrenbach SN, Kelly JCR, Vu C (2004) Integration of child health information systems: current state and local health department efforts. J Public Health Manag Pract (suppl): S30–S35
30. Hinman AR, Eichwald J, Linzer D et al (2005) Integrating child health information systems. Am J Public Health 95:1923–1927
31. Council of State and Territorial Epidemiologists (2004) National assessment of epidemiologic capacity: findings and recommendations. Available at: https://c.ymcdn.com/sites/www.cste.org/resource/resmgr/Workforce/2004ECA.pdf. Accessed 5 Oct 2013
32. Doyle TJ, Glynn MK, Groseclose SL (2002) Completeness of notifiable infectious disease reporting in the United States: an analytical literature review. Am J Epidemiol 155(9):866–874
33. Government Accounting Office. Infectious diseases: gaps remain in surveillance capabilities of state and local agencies. 108th Congress, 1st session, 24 September 2003, p 1
34. http://emergency.cdc.gov/episurv/. Accessed 5 Oct 2013
35. Mandl KD, Overhage JM, Wagner MM, Lober WB, Sebastiani P, Mostashari F et al (2004) Implementing syndromic surveillance: a practical guide informed by the early experience. J Am Med Inform Assoc 11(2):141–150
36. Boulos MNK, Scotch M, Cheung K-H et al (2008) Web GIS in practice VI: a demo playlist of geo-mashups for public health neogeographers. Int J Health Geogr 7:38
37. Beale L, Hodgson S, Abellan JJ, LeFevre S, Jarup L (2010) Evaluation of spatial relationships between health and the environment: the rapid inquiry facility. Environ Health Perspect 118:1306–1312
38. Catelan D, Biggeri A (2010) Multiple testing in disease mapping and descriptive epidemiology. Geospat Health 4(2):219–229
39. Chui KKH, Wenger JB, Cohen SA, Naumova EN (2011) Visual analytics for epidemiologists: understanding the interactions between age, time, and disease with multi-panel graphs. PLoS ONE 6(2):e14683
40. Onicescu G, Hill EG, Lawson AB, Korte JE, Gillespie MB (2010) Joint disease mapping of cervical and male oropharyngeal cancer incidence in Blacks and Whites in South Carolina. Spat Spattemporal Epidemiol 1(2–3):133–141
41. Smieszek T, Balmer M, Hattendorf J, Axhausen KW, Zinsstag J, Scholz RW (2011) Reconstructing the 2003/2004 H3N2 influenza epidemic in Switzerland with a spatially explicit, individual based model. BMC Infect Dis 11:115
42. Tatem AJ, Campiz N, Gething PW, Snow RW, Linard C (2011) The effects of spatial population dataset choice on estimates of population at risk of disease. Popul Health Metrics 9:4
43. Warren-Gash C, Bhaskaran K, Hayward A et al (2011) Circulating influenza virus, climatic factors, and acute myocardial infarction: a time series study in England and Wales and Hong Kong. J Infect Dis 203:1710–1718
44. Yang Y, Atkinson PM, Ettema D (2011) Analysis of CDC social control measures using an agent-based simulation of an influenza epidemic in a city. BMC Infect Dis 11:199
45. Moise IK, Kalipeni E, Zulu LC (2011) Analyzing geographical access to HIV sentinel clinics in relation to other health clinics in Zambia. J Map Geogr Libr 7(3):254–281
46. Short V, Marriott C, Ostroff S et al (2011) Description and evaluation of the 2009–2010 Pennsylvania influenza sentinel school monitoring system. Am J Public Health 101(11):2178–2183
47. Alon D, Shitrit P, Chowers M (2010) Risk behaviors and spectrum of diseases among elderly travelers: a comparison of younger and older adults. J Travel Med 17:250–255
48. Flores-Figueroa J, Okhuysen PC, von Sonnenburg F et al (2011) Patterns of illness in travelers visiting Mexico and Central America: the GeoSentinel experience. Clin Infect Dis 53(6):523–531

49. Freedman DO, Weld LH, Kozarsky PE, Fisk T et al (2006) Spectrum of disease and relation to place of exposure among ill returned travelers. N Engl J Med 354:119–130
50. Han P, Balaban V, Marano C (2010) Travel characteristics and risk-taking attitudes in youths traveling to nonindustrialized countries. J Travel Med 17:316–321
51. Hill DR, Ericsson CD, Pearson RD et al (2006) The practice of travel medicine: guidelines by the Infectious Diseases Society of America. Clin Infect Dis 43:1499–1539
52. da Costa-Ribeiro M, Lourenco-de-Oliveira R, Failloux A (2007) Low gene flow of Aedes aegypti between dengue-endemic and dengue-free areas in southeastern and southern Brazil. AmJTrop Med Hyg 76(2):303–309
53. Lozano-Fuentes S, Fernandez-Salas I, de Lourdes Munoz M et al (2009) The neovolcanic axis is a barrier to gene flow among Aedes aegypti populations in Mexico that differ in vector competence for Dengue 2 virus. PLoS Negl Trop Dis 3:e468
54. Rotela C, Fouque F, Lamfri M et al (2007) Space-time analysis of the dengue outbreak spreading dynamics in the 2004 Tartagal outbreak, Northern Argentina. Acta Trop 103(1):1–13
55. Behrens JJW, Moore CG (2013) Using geographic information systems to analyze the distribution and abundance of Aedes aegypti in Africa: the potential role of human travel in determining the intensity of mosquito infestation. Int J Appl Geospat Res 4(2):9–38
56. Lewis SL, Feighner BH, Loschen WA et al (2011) SAGES: a suite of freely available software tools for electronic disease surveillance in resource-limited settings. PLoS ONE 6(5):e19750
57. Bradley CA, Rolka H, Walker D et al (2005) BioSense: implementation of a national early event detection and situational awareness system. Morb Mortal Wkly Rep 54:11–19
58. Reeder B, Revere D, Olson DR et al (2011) Perceived usefulness of a distributed community-based syndromic surveillance system: a pilot qualitative evaluation study. BMC Res Notes 4:187
59. Hutwagner L, Thompson W, Seeman GM et al (2003) The bioterrorism preparedness and response early aberration reporting system (EARS). J Urban Health Bull NY Acad Med 80:89–96
60. Tsui FC, Espino JU, Dato VM et al (2003) Technical description of RODS: a real-time public health surveillance system. J Am Med Inform Assoc 10:399–408
61. Picciano P, Grant F, Beaubien J et al (2010) BioSINE: an intuitive visualization tool to enhance collaboration between research and practice in disease surveillance. In: Proceedings of the International Society of Disease Surveillance Conference (ISDS 2010), Park City
62. Gardner EM, McLees MP, Steiner JF et al (2011) The spectrum of engagement in HIV care and its relevance to test and- treat strategies for prevention of HIV infection. Clin Infect Dis 52:793–800
63. Fairchild AL, Bayer R (2011) HIV surveillance, public health, and clinical medicine – will the walls come tumbling down? N Engl J Med 365:685–687
64. Department of Health and Human Services (2010) Connections that count. HRSA Care action. Available at: http://hab.hrsa.gov/newspublications/careactionnewsletter/february2010.pdf Accessed 5 Oct 2013.
65. Association of American Geographers, & National Institutes of Health (2011) Report of the AAG-NIH workshop on geospatial infrastructure for medical research. Available at: http://www.aag.org/cs/projects_and_programs/aag_initiative_for_an_nihwide_gis_infrastructure/report_of_the_aagnih_workshop_on_geospatial_infrastructure_for_medical_research_2011. Accessed 5 Oct 2013
66. Elwood S, Goodchild MF, Sui DZ (2012) Researching volunteered geographic information: spatial data, geographic research, and new social practice. Ann Assoc Am Geogr 102(3):571–590
67. McKnight KP, Messina JP, Shortridge AM et al (2011) Using volunteered geographic information to assess the spatial distribution of West Nile Virus in Detroit, Michigan. Int J Appl Geospat Res 2(3):72–85
68. Cook S, Conrad C, Fowlkes AL et al (2011) Assessing Google flu trends performance in the United States during the 2009 influenza virus A (H1N1) pandemic. PLoS ONE 6(8):e23610

Part II
Geographies of Human Health

Chapter 3
Using Geographic Information for Disease Surveillance at Mass Gatherings

Abstract Mass gatherings present the medical community with an excellent window of opportunity to study infectious diseases that can be transmitted over long distances. This is because the venue of a mass gathering usually does not change year-to-year. As a result, special attention can be given to the public health risks that are introduced by travelers from around the world into these mass gatherings. Travelers can also be infected with diseases that are endemic in the host country and transport the locally acquired infectious diseases to their home environments. Therefore, mass gatherings can be thought of as global-to-local-to-global events because of the initial convergence of global populations and the subsequent divergence of populations throughout the world. This chapter discusses three active areas of geographic research that have emerged from our understanding of disease surveillance at mass gatherings: the role of transportation and population geographies in disease surveillance; the spatial and temporal dimensions of environmental geography in the spread of disease; and the advances in GIScience that provide real-world surveillance and monitoring of disease and injuries at mass gatherings.

Keywords Mass gatherings • Disease pandemics • Transportation geography • Population geography • Environmental geography • GIScience

3.1 Introduction

An aspect of disease surveillance that has received renewed attention in recent years is the public health risks associated with mass gatherings [1]. Mass gatherings are defined as public events – organized or spontaneous – that are held for a certain time period and attended by more than 25,000 people. Mass gatherings events can involve sporting events, religious, social, cultural, or political congregations, as well as gatherings of displaced populations due to natural disasters or conflict. Public health risks associated with mass gatherings include infectious disease outbreaks, non-communicable diseases, injuries, environmental exposures (such as heat-related illnesses), and trauma [2–4].

Infectious diseases at mass gatherings can be detected and understood at several levels. For instance, travelers from various global locations can introduce infectious diseases into mass gatherings that can then spread to residents in the local community.

A.J. Blatt, *Health, Science, and Place*, Geotechnologies and the Environment 12,
DOI 10.1007/978-3-319-12003-4_3

Conversely, travelers to mass gatherings can be infected with diseases that are endemic in the local community but not their home countries. These infected individuals – whether originating from the host country or elsewhere – can then transport locally acquire infectious diseases to their home countries, where they might then start new epidemics [5].

Infectious diseases, however, are not the greatest health risks associated with mass gatherings. Non-communicable diseases and injuries have caused more deaths and morbidity than have communicable diseases [6–8]. At the summer Olympic Games in Atlanta, GA, USA, more than 1,000 people received medical care for heat-related illnesses [9]. During the pilgrimage to Mecca, Saudi Arabia, in August 1985, 2,000 cases of heatstroke were reported and more than 1,000 of these individuals died within a few days [6]. In addition, the occurrence of severe acute cardiovascular events are twice as likely during mass gathering events that are associated with intense emotional stress, such as sporting tournaments [10, 11].

Historically, the oldest and largest annual mass gathering is the three-week long Muslim pilgrimage to the Hajj in Saudi Arabia, in which nearly three million pilgrims participate. The pilgrims come from more than 183 countries, which are mostly low-income with weak health systems. In addition to the environment-related health risks described above, these individuals are also likely to have pre-existing health conditions and be more susceptible to infections. This amalgamation of conditions, or "brewing the perfect storm," could lead to an outbreak at the mass gathering, with subsequent spread of infection, upon their return to their home country [12].

The contributions and thought leadership from professional geographers are well-suited to address these issues concerning the understanding, detection, and response to illnesses and disease at mass gatherings. First, since travel is a central and critical aspect of mass gatherings, a robust understanding of the spatial movements and interactions of certain populations is essential to assessing the health risks associated with a mass gathering event. The seasonality and timing of the mass gathering event are also crucial because public health risks are usually correlated with the temporal and spatial distribution of travelers to and from the city where the mass gathering event is hosted. Moreover, environmental and seasonal factors play a clinically relevant role in the infectious activity of pathogens with strong seasonal patterns (i.e., due to the effects of climate). Finally, in countries where public health surveillance and reporting infrastructures are often suboptimal, specialized and focused efforts – such as those on spatial data infrastructure and volunteered geographic information – are vital in the implementation of robust data gathering methods, enhanced analytic capabilities, and improved capacity for electronic disease surveillance [5, 13].

This chapter will expand on each of these ideas by presenting a brief history of disease surveillance at mass gatherings, describing the important health-related risk factors at mass gatherings, and explaining how professional geographers can contribute to the current understanding of disease surveillance at mass gathering events.

3.2 A Brief History of Disease Surveillance at Mass Gatherings

A notable quality of human beings is our willingness to travel long distances to gather in one place, for a variety of reasons – such as religion, politics, sport, and entertainment. Modern modes of transport – such as air, rail, train, and car – have enabled the number of people attending these gathering to increase, as well as the speed at which people travel to these events. Mass gatherings introduce a risk of the spread of infectious diseases, amongst other risks – such as burns, heat exhaustion, dehydration, trauma, human stampedes, and the potential for environmental and public health hazards.

One of the largest single mass gatherings on Earth is the Hindu festival of Kumbh Mela in India. It is believed that this festival likely contributed to the 1817–1824 Asiatic cholera pandemic. Pilgrims are thought to have carried the cholera bacteria from an endemic area in the lower Ganges to populations in the upper Ganges, from there to Kolkata and Mumbai, and across the subcontinent. Then, British soldiers and sailors transported the bacteria to Europe and to the Far East. The epidemic ended very abruptly in 1824, after a harsh and cold winter [14, 15]. In 2013, the same Haridwar-based event attracted approximately 80 million pilgrims between January and April. It is estimated that at least 16 million people were present at the height of the festival on April 14 [16]. Despite much rapid monitoring and public health interventions, diarrheal diseases (such as cholera) continue to be a risk at these mass gatherings today.

The largest annual mass gathering on Earth is the Islamic Hajj pilgrimage to Mecca, Saudi Arabia. During the first Hajj in 632 A.D., the pilgrims reported having a fever-like illness known locally as "Yethrib fever," which is now believed to be malaria. Other major epidemics such as plague and cholera have been reported at the Hajj. In recently years, the number of people attending this event has doubled in the past decade, reaching 3 million in 2012; however, in 2013, concerns over a new SARS-like respiratory virus caused the attendance to decline to 2 million [17]. To place this event in its local context, the influx of pilgrims almost doubles the resident population of Mecca (which is approximately 1.4 million) every year [18]. Because the venue does not change and the event occurs annually, this mass gathering event provides the medical community with an excellent window of opportunity to research infectious diseases that not only affect a mass gathering event but also have the potential for pandemic spread [19, 20].

In recent decades, international sporting events such as the Olympics and the World Cup have attracted global audiences because of affordable air travel and local accommodations. As explained earlier, global attendance and travel can be associated with a heightened risk of imported diseases. For instance, there was a measles outbreak during the 2010 Winter Olympics, in Vancouver, British Columbia, Canada. The infection had spread quickly to the remote areas of British Columbia, causing substantial morbidity among its indigenous peoples [21]. In 2006, an outbreak of chicken pox occurred among members of the Maldives volleyball squad at Doha's Asian Games, but was successfully managed by use of quarantine, antiviral drugs, and vaccine [22].

Instances of respiratory infections at mass gathering events are generally very common. The duration of contact and the amount of shared air are key determinants of the spread of an infection. Influenza is an example of a viral infection that has a short incubation time and can cause both morbidity and mortality at mass gatherings. In July 2008, an outbreak of influenza was reported at the World Youth Day in Sydney, New South Wales, Australia. This outbreak was caused by several strains of influenza viruses, and the spread was expedited by crowded accommodation and low rates of vaccination [23]. Other similar outbreaks have been reported at the 2002 Winter Olympics in Salt Lake City, Utah, USA; and pandemic influenza A H1N1 has been reported at music festivals in Belgium, Serbia, and Hungary [24–27]. Furthermore, the results of a modelling study has suggested that mass gathering events that are held within 10 days before the peak of an influenza epidemic could lead to a 10 % relative increase in infection rates, therefore worsening the outcome for participants [28].

3.3 Important Public Health Risk Factors at Mass Gatherings

Diseases and injuries at mass gathering events can be categorized as communicable or non-communicable. Communicable diseases include infections resulting from a number of different modes of transmission (such as respiratory, food, water, vector, and animal pathogens). Non-communicable diseases are diseases and injuries caused by extreme weather conditions (such as floods, high winds, and high temperatures), acute cardiovascular stress, human stampedes, and acts of terrorism. While most of the public health response at mass gatherings has been on communicable diseases (due to their potentially huge consequences), non-communicable diseases and injuries have actually caused more deaths and morbidity at mass gatherings [29]. The public health response to both groups of diseases is similar to that for public health emergencies or crises in which the existing infrastructure is inadequate for the sudden surge in demand. This section will address the importance of identifying the appropriate environmental, socio-economic, and health characteristics of the mass gathering events and their participants.

The effects of infectious diseases at mass gatherings were first noted for food-borne illnesses, arising from person-to-person transmission. Although the spread of infectious diseases has been reported in relation to the Hajj, the earliest and best modern examples were outbreaks of gastrointestinal illness [30]. Food-borne and water-borne outbreaks of infectious diseases have the potential to spread very rapidly and on a large scale. Factors that influence the spread of infectious diseases include: the types of infections that are endemic to the host country, the types of infections that are endemic to the home countries of the visitors, and the manner in which the populations mix and interact. These factors can be amplified by the international travel and the subsequent dissemination in the home population – which is often faster than the incubation period of almost all infections. Furthermore, the spread of infection and drug-resistant organisms is expedited by the rapid mass movement and mixing of infectious and susceptible populations [14]. The infections that are likely to arise are also partly determined by the type of event – for instance, events with overcrowded accommodations will be more susceptible to outbreaks of respiratory infections, where as those with excessive consumption of recreational drugs and alcohol could increase the likelihood of sexually transmitted infections [31].

Since respiratory transmissions of infectious disease require close proximity between the infectious agent and host, this is the most common type of infection at events with overcrowded accommodations. The dynamics of transmission are determined by: incubation time, whether the infection is transmitted through droplets or is airborne, and nature of the infectious agent. Examples of respiratory infections are: influenza and other respiratory viruses, tuberculosis, measles, mumps, and meningococcal meningitis. Tuberculosis is a respiratory infection with a long incubation time. Because tuberculosis infections can remain latent for months to years, transmissions are often not noticed during mass gathering events. Although transmission has been reported during air travel, there is currently insufficient evidence to conclude that air travel is an important source of tuberculosis infection [32].

Furthermore, vector-borne diseases (such as malaria, dengue, West Nile encephalitis, and yellow fever) can cause outbreaks in countries where they are not endemic if a traveler is infected and an appropriate vector is present in the host country. Moreover, infectious diseases caused by known human pathogens or occasionally emerging infections can be transmitted from animals to people. For instance, in 1998, an outbreak of leptospirosis was reported among triathlon athletes in Springfield, Illinois, USA, where 11 % of tested participants were positive for the infection. This large outbreak likely arose from athletes drinking contaminated lake water, and demonstrated that unusual infections can occur among those who come into contact with fresh water during mass gathering events in temperate countries [33].

Weather and other environmental conditions, such as warm and cold temperatures, precipitation, and pollution, can contribute to the incidence of non-communicable illnesses, including life-threatening heatstroke, hypothermia, trauma, and dyspnea. A review of the effects of warm weather showed a strong correlation between high temperatures or humidity and the use of medical care during mass gathering events [34]. The heat index, a variable that combines measured air temperature and humidity, was found to be an important predictor of the demand for medical care [35]. Studies in Mecca have found that extreme environmental temperatures led to a major disaster – more than 1,000 deaths from heatstroke in just a few days [6]. In addition, cold temperatures, floods, high winds, and air pollution have been linked to health problems at mass gathering events as well [36–40].

3.4 Using Geographic Intelligence to Improve Disease Surveillance at Mass Gatherings

As demonstrated above, mass gatherings can exacerbate public health risks, on a global level, because travelers from around the world can introduce infectious diseases into mass gatherings, and these infectious diseases can then spread to other people in the host country. Travelers can also be infected with diseases that are endemic in the host country and can then transport locally acquired infectious diseases to their home environments, where they might then start new epidemics. Therefore, mass gatherings can be conceptualized as global-to-local-to-global events because of the initial convergence of global populations and the subsequent divergence of populations throughout the world. Three active areas of geographic research have emerged in our modern-day understanding of disease surveillance at mass gatherings: transportation and population geographies and their role in disease surveillance; the spatial and temporal dimensions of environmental geography in the spread of disease; and the ability of advances in GIScience to provide real-world surveillance and monitoring of disease and injuries.

3.4.1 Transportation and Population Geographies: Their Role in Disease Surveillance

Geographers have long observed that the mobility (of humans, of insect and animal vectors, and of pathogens) plays a prominent role in determining the observable patterns of disease outcomes [41–45]. Advances in modern-day commercial travel –by air, land, and water – have profoundly transformed the global movement of populations throughout the world. More than two billion people travel on commercial air flights every year, and this phenomenon presents new opportunities for locally emerging infectious diseases to quickly transform into international epidemics or pandemics [46, 47]. The accessibility of international modes of transportation has also contributed to an increase in the overall number, frequency, and scale of international mass gathering events. Since transportation and migration are central features of mass gatherings, transportation and population geographers can play a unique and essential role to our understanding of the spatial and temporal aspects of global outbreaks related to mass gatherings [48–52].

Mass gathering events often involve one of four major population groups: participants, observers, residents, and bystanders. Participants are those directly participating in the events, such as pilgrims at a religious festival. Observers are spectators of an event. Residents are those in the host country who interact with participants at the mass gathering – such as food vendors and hotel staff. Bystanders are those who travel to the host country for reasons unrelated to the mass gathering, but interact with people in the mass gathering environment. Since the global origins, mobility, and infectious disease burdens differ for each population group, understanding each group's daily mobility and transportation patterns is essential to assessing the public health risks at a mass gathering event [5].

Modelling the mobility of travelers from their global origins to a mass gathering event is complicated by the fact that the prior travel experiences to mass gatherings events in different locations and during different seasons may not be readily transferrable to future mass gathering events [53]. Considerations must also be given to the pre-departure health risks of each population group, their accommodations at the host country, and their interactions at the mass gathering event. For example, participants (i.e., political leaders and their entourages) of the G20 Summit would be expected to have low pre-travel risks of infectious disease, reside in privileged environments where food or water-borne diseases are unlikely, and be highly sequestered from other population groups at the mass gathering event. Conversely, pilgrims traveling to a religious mass gathering, such as the Kumbh Mela, may have pre-existing health issues, be living in environments where the risks of acquiring food or water-borne diseases are increased, and have extensive interactions with other population groups [5].

3.4.2 *Understanding Disease Ecology's Contributions to Disease Surveillance*

By necessity, disease surveillance at mass gathering events must take into account the environmental and social conditions that promote or sustain illness. Disease ecology is the branch of medical geography that is concerned with the integration of the environmental and social aspects of human lives into an understanding of diseases and injuries [54–57]. Since diseases obviously do not exist independently of environments or hosts, it is important to understand how human behavior, with its cultural and socioeconomic contexts, interacts with the environment to promote or prevent disease [58]. Anticipated (or unanticipated) disturbances in the environment – such as climate change, population growth, urbanization, and migration – can have either positive or negative effects on the incidence or spread of disease [59].

Medical geographer Dr. Melinda Meade recognized that human disease is the outcome of a complex and dynamic interaction between the internal and external environments of an individual or a population. She proposed a view of disease as being an intersection of three types of variables: population, environment, and behavior (Fig. 3.1). Population variables are factors that affect individuals' responses to disease as biological beings, such as nutritional and immunological status, age, and gender. Environmental variables are all aspects of the built, natural, and social environments that can affect disease outcomes. Behavior variables include both observable aspects of actions and culture, such as social organization, technology, diet, as well as less tangible variables like perceptions of risk. Disease outcomes are the result of place- and time- specific interactions among these variables [60, 61].

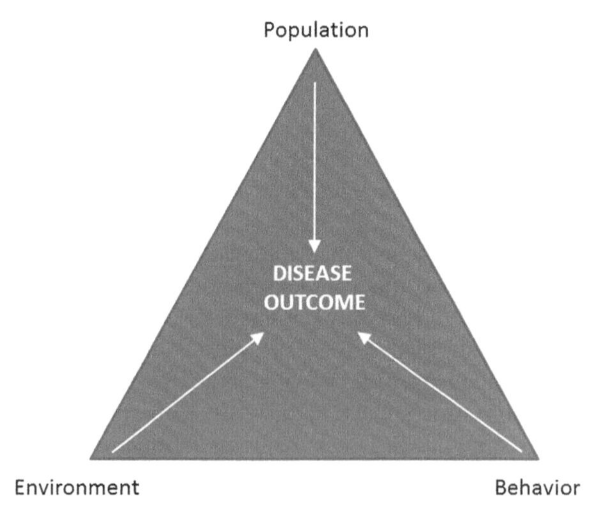

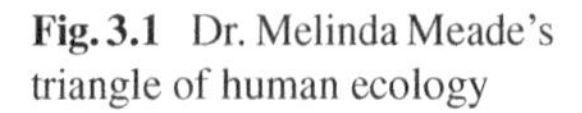
Fig. 3.1 Dr. Melinda Meade's triangle of human ecology

Understanding the relationship between disease ecology and disease surveillance is an essential component of preventing illness at mass gathering events because this knowledge can help health officials understand how pathogens cause disease in people and what factors affect the severity of these illnesses. Environmental factors, such as patterns of precipitation and minimum mean temperature, often influence the distribution and abundance of disease vectors worldwide. Without sufficient rain and temperature, disease vector communities cannot sustain themselves over time. Competition for resources in a drier climate can also lead to changes in vector community composition. These abiotic and site specific ecological factors determine where endemic disease is and is not sustainable [62–64]. Geographers play a crucial leadership role in identifying and filling in the current knowledge gaps about diseases, ranging from basic scientific research to the societal effects of infections and non-communicable illnesses. This information is important in modern public health and policy developments related to the planning for mass gathering events in the future [65].

3.4.3 GIScience and the Spatial Turn in Disease Surveillance

Geographic information science (GIScience) has a strong history of providing support to traditional disease surveillance. Many electronic disease surveillance systems have a mapping module to allow public health officials to visualize disease outbreaks on the map [66–69]. In addition, geographic information systems (GIS) provide a baseline for monitoring and evaluating outbreak investigations at mass gathering events, so that the geographic progression of disease is continually monitored. Public health field personnel use mobile GIS and global position system (GPS) devices to more efficiently navigate to locations for data collection.

Maps, imagery, and descriptive metadata are used to identify high-transmission areas or areas with environmental conditions ideal for disease vectors [45, 48, 50–52]. In short, GIS provides public health officials with the capabilities to spatially visualize and analyze complex spatio-temporal relationships between the infectious agent, host, and the environment at mass gathering events.

In addition, volunteered geographic information (VGI) has been shown to facilitate and improve public health reporting and communications at mass gathering events. Data collection and reporting of public health intelligence during a mass gathering event have been challenging tasks because of the large numbers of attendees, the rapid and large-scale movements of people, and their fairly short stay at the mass gathering event. Deploying VGI-based technology at mass gathering events is an effective way of gathering and disseminating public health information (as more than 90 % of the global population is covered by mobile phone networks and more than 140 countries provide third-generation wireless telecommunication services) [70–74].

An example of a successful deployment of VGI-based public health intelligence is during the 2009 Hajj, by the Saudi Arabian government. The goals were to use smart devices to assist in the early detection of emerging outbreaks of infectious diseases, and to improve the efficiency of case reporting and operational effectiveness, by using meaningful data visualizations and geo-statistical analyses with geographically tagged data (Fig. 3.2). This project monitored nine infectious diseases: pandemic influenza A H1N1, influenza-like illness, meningococcal meningitis, viral hemorrhagic fever, plague, yellow fever, cholera, foodborne illness, and poliomyelitis. Questionnaires about these diseases were uploaded to a central server, and wirelessly disseminated to laptops and smart devices in Saudi Arabia at different points of data gathering. Field investigators were assigned to local clinics and

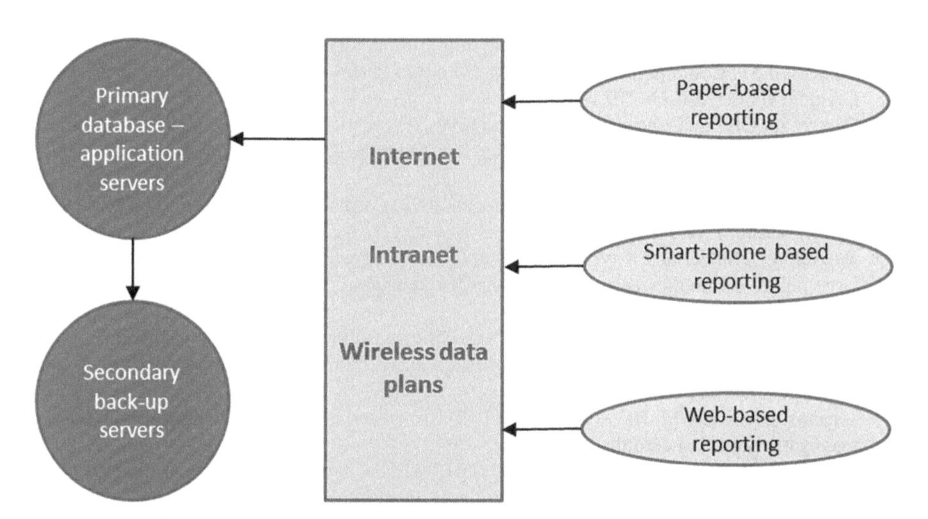

Fig. 3.2 Information technology infrastructure used for public health surveillance during the 2009 Hajj, Mecca, Saudi Arabia (Adapted from Khan et al. [12])

hospitals in and around Mecca. Real-time information from these investigators was quickly analyzed and synthesized by epidemiologists and presented to public health officials in the form of daily reports.

This project demonstrated that VGI-based smart technology could be swiftly and successfully integrated into the existing information technology infrastructure of Saudi Arabia's Ministry of Health. Wireless telecommunications provided ample opportunities for field investigators to gather geographically tagged health information electronically, at the point of contact with the pilgrims at the mass gathering event. This efficient real-time transfer of information formed the basis of an integrated platform used to synthesize epidemiological, clinical, and laboratory information into actionable public health intelligence for this mass gathering event [5].

References

1. McConnell J (2012) Mass gatherings health series. Lancet Infect Dis 12:8
2. Kutsar K (2012) Public health preparedness in mass gatherings. EpiNorth 13:88–90
3. Lombardo JS, Sniegoski CA, Loschen WA et al (2008) Public health surveillance for mass gatherings. Johns Hopkins APL Tech Dig 27(4):347–355
4. Polkinghorne BG, Massey PD, Durrheim DN et al (2013) Prevention and surveillance of public health risks during extended mass gatherings in rural areas: the experience of the Tamworth Country Music Festival. Aust Public Health 127:32–38
5. Khan K, McNabb SJN, Memesh ZA et al (2012) Infectious disease surveillance and modelling across geographic frontiers and scientific specialties. Lancet Infect Dis. doi:10.1016/S1473-3099(11)70313-9
6. Ghaznawi HI, Ibrahim MA (1987) Heatstroke and heat exhaustion in pilgrims performing the Haj (annual pilgrimage) in Saudi Arabia. Ann Saudi Med 7:323–326
7. Hanslik T, Boelle PY, Flahault A (2001) Setting up a specific surveillance system of community health during mass gatherings. J Epidemiol Community Health 55:683–684
8. Wetterhall SF, Coulombier DM, Herndon JM et al (1998) Medical care delivery at the 1996 Olympic Games. JAMA 279:1463–1468
9. Centers for Disease Control and Prevention (1996) Prevention and management of heat-related illness among spectators and staff during the Olympic Games-Atlanta, July 6–23, 1996. MMWR Wkly 45:631–633
10. Wilbert-Lampen U, Leistner D, Greven S et al (2008) Cardiovascular events during World Cup soccer. N Engl J Med 358:475–483
11. Weaver WD, Sutherland K, Wirkus MJ et al (1989) Emergency medical care requirements for large public assemblies and a new strategy for managing cardiac arrest in this setting. Ann Emerg Med 18:155–160
12. Ahmed QA, Barbeschi M, Memish ZA (2009) The quest for public health security at Hajj: the WHO guidelines on communicable disease alert and response during mass gatherings. Travel Med Infect Dis 7:226–230
13. Johannson A, Batty M, Hayashi K et al (2012) Crowd and environmental management during mass gatherings. Lancet Infect Dis 12:150–156
14. Abubakar I, Gautret P, Brunette GW et al (2012) Global perspectives for prevention of infectious diseases associated with mass gatherings. Lancet Infect Dis 12:66–74
15. Hays JN (2005) Epidemics and pandemics: their impacts on human history. ABC-CLIO, Santa Barbara, pp 214–219

16. Spinney L (2013) At largest religious festival, some abandon elderly. National Geographic. Available at http://news.nationalgeographic.com/news/2013/02/130223-culture-travel-religion-india-maha-kumbh-mela-world-women-widow-hindu-festival. Accessed 23 Dec 2013
17. Haridi N (2013) Hajj Virus concerns reduce Mecca Muslims Pilgrims by one million. The Huffington Post. Available at http://www.huffingtonpost.com/2013/10/13/hajj-virus-_n_4094644.html. Accessed 23 Dec 2013
18. Memish ZA, Stephens GM, Steffen R et al (2012) Emergence of medicine for mass gatherings: lessons from the Hajj. Lancet Infect Dis 12:56–65
19. Farid MA (1956) Implications of the Mecca pilgrimage for a regional malaria eradication programme. Bull World Health Organ 15:828–833
20. Ahmed QA, Yaseen MA, Memish ZA (2006) Health risks at the Hajj. Lancet 367:1008–1015
21. ProMED Mail. Measles–Canada: (BC), imported. Available at http://www.promedmail.org/direct.php?id=20100406.1102. Accessed 23 Dec 2013
22. ProMED Mail. Varicella, Asian Games–Qatar ex Maldives. Available at http://www.promedmail.org/direct.php?id=20061129.3385. Accessed 23 Dec 2013
23. Blyth CC, Foo H, van Hal SJ et al (2010) Influenza outbreaks during World Youth Day 2008 mass gathering. Emerg Infect Dis 16:809–815
24. Gundlapalli AV, Rubin MA, Samore MH et al (2006) Influenza, Winter Olympiad, 2002. Emerg Infect Dis 12:144–146
25. Gutiérrez I, Litzroth A, Hammadi S et al (2009) Community transmission of influenza A (H1N1) virus at a rock festival in Belgium, 2–5 July 2009. Euro Surveill 14:19294
26. Loncarevic G, Payne L, Kon P et al (2009) Public health preparedness for two mass gathering events in the context of pandemic influenza (H1N1) 2009—Serbia, July 2009. Euro Surveill 14:19296
27. Botelho-Nevers E, Gautret P, Benarous L et al (2010) Travel-related influenza A/H1N1 infection at a rock festival in Hungary: one virus may hide another one. J Travel Med 17:197–198
28. Shi P, Keskinocak P, Swann JL et al (2010) The impact of mass gatherings and holiday traveling on the course of an influenza pandemic: a computational model. BMC Public Health 10:778
29. Steffen R, Bouchama A, Johansson A et al (2012) Non-communicable health risks during mass gatherings. Lancet Infect Dis 12:142–149
30. Gross M (1976) Oswego County revisited. Public Health Rep 91:160–170
31. Lim MS, Hellard ME, Aitken CK et al (2007) Sexual-risk behaviour, self-perceived risk and knowledge of sexually transmissible infections among young Australians attending a music festival. Sex Health 4:51–56
32. Abubakar I (2010) Tuberculosis and air travel: a systematic review and analysis of policy. Lancet Infect Dis 10:176–183
33. Morgan J, Bornstein SL, Karpati AM et al (2008) Outbreak of leptospirosis among triathlon participants and community residents in Springfield, Illinois, 1998. Clin Infect Dis 34:1593–1599
34. Baird MB, O'Connor RE, Williamson AL et al (2010) The impact of warm weather on mass event medical need: a review of the literature. Am J Emerg Med 28:224–229
35. Perron AD, Brady WJ, Custalow CB et al (2005) Association of heat index and patient volume at a mass gathering event. Prehosp Emerg Care 9:49–52
36. Yazawa K, Kamijo Y, Sakai R et al (2007) Medical care for a mass gathering: the Suwa Onbashira Festival. Prehosp Disaster Med 22:431–435
37. Li Y, Wang W, Kan H et al (2010) Air quality and outpatient visits for asthma in adults during the 2008 Summer Olympic Games in Beijing. Sci Total Environ 408:1226–1227
38. Schulte D, Meade DM (1993) The papal chase. The Pope's visit: a "mass" gathering. Emerg Med Serv 22:46–9, 65–75, 79
39. Brunko M (1989) Emergency physicians and special events. J Emerg Med 7:405
40. Dress JM, Horton EH, Florida R (1995) Music, mud & medicine. Woodstock '94: a maniacal, musical mass-casualty incident. Med Serv 24:21, 30–32
41. Roundy RW (1978) A model for combining human behavior and disease ecology to assess disease hazard in a community: rural Ethiopia as a model. Soc Sci Med Part D: Med Geogr 12:121–130

42. Meade MS (1977) Medical geography as human ecology: the dimension of population movement. Geogr Rev 67:379–393
43. Prothero RM (1961) Population movements and problems of malaria eradication in Africa. Bull World Health Organ 24:405
44. Prothero RM (1963) Population mobility and trypanosomiasis in Africa. Bull World Health Organ 28:615
45. Behrens JJW, Moore CG (2013) Using geographic information systems to analyze the distribution and abundance of Aedes aegypti in Africa: the potential role of human travel in determining the intensity of mosquito infestation. Int J Appl Geospat Res 4(2):9–38
46. Gushulak BD, MacPherson DW (2000) Population mobility and infectious diseases: the diminishing impact of classical infectious diseases and new approaches for the 21st century. Clin Infect Dis 31:776–780
47. WHO (2007) The World Health Report 2007–a safer future: global public health security in the 21st century. Available at http://www.who.int/whr/2007/en/index.html. Accessed 26 Dec 2013
48. Wen T, Lin M, Fang C (2012) Population movement and vector-borne disease transmission: differentiating spatial–temporal diffusion patterns of commuting and noncommuting dengue cases. Ann Assoc Am Geogr 102:1026–1037. doi:10.1080/00045608.2012.671130
49. Kan CC, Lee PF, Wen TH et al (2008) Two clustering diffusion patterns identified from the 2001–2003 dengue epidemic, Kaohsiung, Taiwan. Am J Trop Med Hyg 79:344–352
50. Adams B, Kapan DD (2009) Man bites mosquito: understanding the contribution of human movement to vector-borne disease dynamics. PLoS ONE 4:e6763
51. Stoddard ST, Morrison AC, Vazquez-Prokopec GM et al (2009) The role of human movement in the transmission of vector-borne pathogens. PLoS Negl Trop Dis 3:e481
52. Vazquez-Prokopec GM, Stoddard ST, Paz-Soldan V et al (2009) Usefulness of commercially available GPS data-loggers for tracking human movement and exposure to dengue virus. Int J Health Geogr 8:68
53. Khan K, Freifeld CC, Wang J et al (2010) Preparing for infectious disease threats at mass gatherings: the case of the Vancouver 2010 Olympic Winter Games. CMAJ 182:579–583
54. Dubos RJ (1987) Mirage of health: utopias, progress and biological change. Rutgers University Press, New Brunswick
55. Thammapalo S, Chongsuvivatwong V, Geater A et al (2008) Environmental factors and incidence of dengue fever and dengue haemorrhagic fever in an urban area. South Thail Epidemiol Infect 136:135–143
56. May JM (1950) Medical geography: its methods and objectives. Geogr Rev 40:9–41
57. Hunter JM (1974) The challenge of medical geography. In: Hunter JM (ed) The geography of health & disease. Department of Geography, University of North Carolina, Chapel Hill, pp 1–31
58. Meade MS, Emch M (2010) Medical geography. Guilford, New York, p 26
59. Mayer JD (2000) Geography, ecology and emerging infectious diseases. Soc Sci Med 50:937–952
60. Carrel M, Emch M (2013) Genetics: a new landscape for medical geography. Ann Assoc Am Geogr. doi:10.1080/00045608.2013.784102
61. Blatt AJ (2013) Geospatial applications in disease surveillance: solutions for the future. Int J Appl Geospat Res 4(2):1–8
62. Aiken R, Leigh C (1978) Dengue hemorrhagic fever in south-east Asia. Trans Inst Br Geogr 3(4):476–497. doi:10.2307/622125
63. Kovats R, Campbell-Lendrum D, McMichael A et al (2001) Early effects of climate change: do they include changes in vector borne disease? Philos Trans R Soc Lond B Biol Sci 356(1411):1057–1068. doi:10.1098/rstb.2001.0894, PMID: 11516383
64. Reiter P (2001) Climate change and mosquito borne disease. Environ Health Perspect 109(1):141–161. doi:10.1289/ehp.01109s1141, PMID: 11250812
65. Tam JS, Barbeschi M, Shapovalova N et al (2012) Research agenda for mass gatherings: a call to action. Lancet Infect Dis 12:231–239

66. Lewis SL, Feighner BH, Loschen WA et al (2011) SAGES: a suite of freely available software tools for electronic disease surveillance in resource-limited settings. PLoS ONE 6(5):e19750
67. Hutwagner L, Thompson W, Seeman GM et al (2003) The bioterrorism preparedness and response early aberration reporting system (EARS). J Urban Health Bull NY Acad Med 80:89–96
68. Tsui FC, Espino JU, Dato VM et al (2003) Technical description of RODS: a real-time public health surveillance system. J Am Med Inform Assoc 10:399–408
69. Picciano P, Grant F, Beaubien J et al (2010) BioSINE: an intuitive visualization tool to enhance collaboration between research and practice in disease surveillance. In: Proceedings of the International Society of Disease Surveillance Conference (ISDS 2010), Park City
70. Aanensen DM, Huntley DM, Feil EJ et al (2009) EpiCollect: linking smartphones to web applications for epidemiology, ecology and community data collection. PLoS ONE 4:e6968
71. International Telecommunication Union (2013) World telecommunication/ICT indicators database. Available at http://www.itu.int/en/ITU-D/Statistics/Pages/stat/default.aspx. Accessed 27 Dec 2013
72. Hoffman JA, Cunningham JR, Suleh AJ et al (2010) Mobile direct observation treatment for tuberculosis patients: a technical feasibility pilot using mobile phones in Nairobi. Kenya Am J Prev Med 39:78–80
73. Safaie A, Mousavi SM, LaPorte RE et al (2006) Introducing a model for communicable diseases surveillance: cell phone surveillance (CPS). Eur J Epidemiol 21:627–632
74. Yu P, de Court PE, Galea G et al (2009) The development and evaluation of a PDA-based method for public health surveillance data collection in developing countries. Int J Med Inform 78:532–542

Chapter 4
Geographical Representations of Patients and Their Health Conditions

Abstract This chapter briefly reviews the well-established paradigms in patient representations of health. Over the last 50 years, the view of the patient has evolved from a person affected by an infectious disease to a person who is constantly changing and adapting to changes in a social and natural environment of hosts, agents, and reservoirs. Extensive research has demonstrated that the quality of neighborhoods (i.e., their environmental and social contexts) matters to human health. The chapter moves on to examine how viewing a patient through the lens of a landscape, as an organizing principle, can improve patient care and delivery. For instance, when a patient's medical history is combined with his or her place history in an electronic medical record (EMR), what additional insights can a physician gain to improve the quality and timeliness of a clinical diagnosis? Finally, the chapter concludes with a discussion of how combining qualitative and quantitative GIS methodologies can uncover the health conditions of minority populations more readily, such that life-saving treatments can commence before problematic symptoms appear.

Keywords Patient representations of health • Social contexts • Disease ecology • Place history • Qualitative GIS

4.1 Introduction

As stated in several places along our discussion, one of the compelling forces behind this book is the current attention on health care reform in the United States. The Patient Protection and Affordable Care Act (PPACA) of 2010 provided funding and legislation for a fully operationalized national health information exchange (HIE) that is designed improve patient care coordination. It achieves this goal by allowing physicians to share HIPAA-compliant medical records with the network of providers caring for a patient (Chap. 1, Fig. 1.3).

This renewed attention on improving health outcomes through the sharing of electronic health information provides the geographic community with an unprecedented opportunity to communicate how geographical thought and knowledge can contribute to our understanding of a patient's medical condition and personal history. For instance, answers to the following questions can now be

A.J. Blatt, *Health, Science, and Place*, Geotechnologies and the Environment 12,
DOI 10.1007/978-3-319-12003-4_4

addressed: How can viewing a patient through the lens of a landscape, as an organizing principle, improve patient care and delivery? When a patient's medical history is combined with his or her place history in an electronic medical record (EMR), what additional insights can a physician gain to improve the quality and timeliness of a clinical diagnosis? How can the health conditions of minority populations be uncovered more readily, so that treatments can commence before problematic and life-threatening symptoms appear?

This chapter will address the importance of these geographically relevant topics juxtaposed with our current view of patients and their health conditions. First, it will briefly review some of the established paradigms in patient representations – such as masculine versus feminist perspectives, the environment impacts on health and wellness, and the role of social segregation in population health outcomes. Next, the discussion will move on to discuss the important role of the landscape (both social and environmental) as a unifying and organizing principle with which to view a patient's medical conditions. Finally, this chapter will conclude with a discussion of the utility of geographic information systems (GIS) technology in qualitative and quantitative approaches to disease surveillance, as well as in the detection and understanding of the underserved populations that are often overlooked by the mainstream medical community today.

4.2 Representations of Patients and Their Health Conditions

Physicians and the medical community have long described patients and their health conditions along three main dimensions: gender, age, and race. Indeed, a number of geographic and epidemiological studies have examined the differences between men and women, in terms of disease prevalence and risk factors [1–5]. Researchers have also focused on the wide-ranging effects of age on health and disease etiology [6–8]. Furthermore, race has been well-established as an important risk factor in many acute and chronic conditions, such as hypertension and diabetes [9–12].

In addition to these three dimensions, researchers have also recognized that a person's socio-economic status and neighborhood of residence are two important determinants of human health and quality of life. It is generally accepted that a person's quality of life varies according to his or her neighborhood of residence. For instance, some neighborhoods can have characteristics that enhance the residents' quality of life, while other neighborhoods have characteristics that have been shown to be harmful [13–18]. Emerging from these observations is the idea of a neighborhood effect, which is the combined effect of a neighborhood and its composite characteristics on individual- and population-level health outcomes [19–21].

Most neighborhood effects studies have been conducted within the context of race and the socioeconomic characteristics of different places [22, 23]. There are four main approaches to studying neighborhood effects. The first approach relies on census demographic and socioeconomic data to show how certain structural neighborhood characteristics – such as low income and high unemployment rates – erode human health and increase the risk of illness and disease. The second approach focuses on the social and institutional processes that are related to problem behaviors (e.g., ranging from teenage pregnancy to dropping out of school) resulting in adverse outcomes [24, 25]. The third approach involves a systematic social observation of neighborhoods to record activities related to physical or social disorders [26]. The fourth approach involves space–time analyses, which incorporates where a population resides, as well as the location of daily activities such as traveling to work or to shop [27, 28].

The importance of neighborhood economic status on health outcomes varies from country to country [29]. For instance, unlike the United States, many industrialized countries with universal health care display a diminished effect of neighborhood economic status on health [30]. However, even in two countries with universal health care, health outcomes can differ greatly, if socioeconomic segregation and deprivation are significantly more pronounced in one country than another country [29].

Extensive scientific research has demonstrated that the quality of neighborhoods (i.e., their environmental and social contexts) matters to human health. For instance, residents living in neighborhoods characterized by a low quality social and physical environment are associated with a greater likelihood of suffering from depression [31]. In addition, minority neighborhoods in concentrated poverty areas were found to have more than twice the level of traffic density, and its residents are disproportionately exposed to vehicle-related pollutants with serious health consequences [32]. As a result, these neighborhood residents have a higher risk of experiencing the adverse health effects associated with vehicle-related pollutants, such as adverse birth outcomes, respiratory illnesses, cancer, and mortality [33–36].

4.3 Seeing Patients and Their Health Conditions Through the "Lens" of a Landscape

Over the last 50 years, the view of a patient has evolved from a person affected by an infectious disease to a person who is constantly changing and adapting to changes in a social and natural environment of hosts, agents, and reservoirs. Dr. Jacques May, a medical doctor who is considered the "father" of American medical geography, observed that disease is a "multifactorial phenomenon which occurs only if factors coincide in space and time" [37, 38]. He viewed medical geography as the study of the geographical environments and culture landscapes in which disease agents, reservoirs, and hosts exist. Indeed, many studies have found that cultural choices (such as house type, diet, crop preference, and clothing) play an important role in exposing populations to certain disease conditions [39, 40]. Moreover, as people continue to alter the natural environment in ways that affect disease patterns, linking biogeography with human geography can allow medical geographers to explore the relationship between humans and their external environments [40].

The fundamental idea behind disease ecology is that human life is an evolving process that is a "landscape" of interactions between its internal and external environments [41–43]. A relatively new field that emerged in the second half of the twentieth century, disease ecology was a reaction to the belief in the medical community that infectious diseases were a thing of the past and that the ability to cure disease laid in prescribing the correct medication. Disease ecologists believe that, in order to understand a disease, one must understand the *landscape* (i.e., both the person and place) in which the infection occurs [43]. Disease ecologists are also concerned with how humans adapt their behaviors or modify their environments in reaction to changes elsewhere in the environment [44, 45]. Because human hosts, pathogens, and insect vectors are constantly adapting to new and changing conditions (such developing resistance to drugs, and buffers to disease exposure), disturbances in the environment – through climate change, population growth, urbanization, agriculture and migration – can have positive or deleterious effects on the spread of disease, by magnifying or minimizing risk and exposure. As such, disease ecologists often go beyond the practice of describing the spatial patterns of disease and risk factors, and study the process and character of disease outcomes. In disease ecology, geographic variations do not merely drive variation in disease outcomes; geographic variations also create differential rates of change in pathogens and their molecular characteristics (like drug resistance) across space and time [46, 47].

Therefore, it is now possible to push beyond the traditional questions concerning the presence–absence disease outcomes and address issues about which features of human–natural landscapes drive disease emergence and pathogenic evolution. For instance, what behavioral or natural environmental features act as barriers to the spread of drug-resistant pathogens? What population-environment interactions drive diffusion of new disease variants across a landscape? How do spatially variable interventions, such as vaccination or vector eradication, impact pathogen genetic distributions?

Although the majority of public health and medical geography studies of infectious disease still treat the diseases themselves as static occurrences, disease can most accurately be described as the result of a maladaptation between humans and their environment. Linking medical geography and landscape genetics now provides a view of disease that reveals how the population and environmental processes allow disease to persist and evolve. This can aid public health and medical researchers in identifying landscapes where diseases cannot persist, and in providing insight into human-environmental interactions beneficial to human hosts and detrimental to pathogenic success. Consequently, as people continue to modify their environments in a myriad of ways – through increased human mobility, changing water and hygiene interactions, population growth, and the intensification of agricultural production – the fieldwork of medical geographers working at multiple spatial scales to uncover both seen and unseen landscape-level processes will be vital to disease surveillance and health care services delivery.

4.4 Understanding Minority Populations and Hidden Conditions with Qualitative and Quantitative GIS

Although qualitative methods have been used widely in geographic research, human geographers have witnessed considerable advances in qualitative methodologies in recent years [48–50]. Qualitative research refers to research that produces findings not derived from statistical or quantitative procedures. It includes research about people's lives, experiences, behaviors, emotions, and feelings that uses information-intensive methods such as in-depth interviews, focus groups, and participant observation. For geographers, qualitative research seeks to understand and interpret the human experience in its socio-spatial contexts [51, 52].

Qualitative research is particularly helpful for understanding the experience of people with health problems, such as chronic illness or drug addiction, as it attempts to obtain the details about complex phenomena, such as feelings and emotions, which are often difficult to understand through other methods. Qualitative research is also useful for uncovering the silenced voices of marginalized individuals and social groups whose feelings and thoughts have been ignored by the dominant discourses of powerful groups in contemporary society. For instance, the experiences of women, ethnic minorities, poor people, and children are often omitted in official historical records. The use of a variety of archival sources and qualitative materials can help recover the stories and everyday experiences of these people [53–55].

As geographic information systems (GIS) technologies are being widely used by many as a tool for the storage and analysis of quantitative data, their use in qualitative or mixed-method research has been growing [56–58]. In addition to using GIS data and analyses to complement, verify, and interpret the knowledge acquired from qualitative data, geographers also use GIS technologies to integrate the analysis of quantitative and qualitative data [59–64]. For example, in a multi-site study of low-income and welfare-recipient families and their children, family ethnographic field-notes are linked with neighborhood-level census data. The integration of GIS and ethnography has allowed researchers of the project to visualize and better understand the complexity of the lives of low-income families and the strategies they adopt in negotiating the welfare system [61]. Another study in Philadelphia, Pennsylvania has demonstrated the importance of qualitative GIS by capturing the activity space characteristics, perceptions of places, and social network data for investigating substance use, health outcomes, and crime behaviors [8].

Although it is possible to collect high-resolution data on people's daily exercise activities and trips to health care facilities, using location-aware devices like global positioning systems (GPS) and mobile phones, high-quality data are still costly and time consuming to collect [65–68]. Nevertheless, qualitative GIS still holds much promise for the analysis of people's spatiotemporal experiences and the integration of data and analyses that are used to view patients and their health conditions through the lens of a "landscape" [60, 64].

In conclusion, geographic context is an important idea in public health research. It is a foundational concept in many qualitative and quantitative studies that assess people's exposure to contextual or environmental influences. Static conceptualizations of the geographic context (without incorporating the temporal dimension) are problematic because residential location is only one of the many places where people spend their time. Much of the contextual or environmental influence they experience and most of the physical and social resources they utilize might be located outside or far from their residential neighborhoods. The next section in this volume will describe the technological advances (like GPS and Web 2.0) and challenges in collecting, storing, and managing the vast troves of health information about diverse populations and their place histories.

References

1. Courtenay WH (2000) Constructions of masculinity and their influence on men's well-being: a theory of gender and health. Soc Sci Med 50(10):1385–1401
2. Craddock S, Brown T (2009) Representing the un/healthy body. In: Brown T, McLafferty S, Moon G (eds) A companion to health and medical geography. Wiley-Blackwell, Oxford, pp 301–321
3. Fisher JD, Smith LR, Lenz EM (2010) Secondary prevention of HIV in the United States: past, current, and future perspectives. J Acquir Immune Defic Syndr 55:S106–S115
4. Mathew R, Gucciardi E, De Melo M et al (2012) Self-management experiences among men and women with type 2 diabetes mellitus: a qualitative analysis. BMC Fam Pract 13:122
5. Grimaud O, Lapostolle A, Berr C et al (2013) Gender differences in the association between socioeconomic status and subclinical atherosclerosis. PLoS ONE 8(11):e80195. doi:10.1371/journal.pone.0080195
6. Landrigan PJ, Schechter CB, Lipton JM et al (2002) Environmental pollutants and disease in American children: estimates of morbidity, mortality, and costs for lead poisoning, asthma, cancer, and developmental disabilities. Environ Health Perspect 110(7):721–728
7. Cotterell J (2007) Social networks in youth & adolescents. Routledge, London/New York
8. Mennis J, Mason MJ (2011) People, places, and adolescent substance use: integrating activity space and social network data for analyzing health behavior. Ann Assoc Am Geogr 101(2):272–291. doi:10.1080/00045608.2010.534712
9. Bell J, Zimmerman F, Almgren G et al (2006) Birth outcomes among urban African American women: a multilevel analysis of the role of racial residential segregation. Soc Sci Med 63:3030–3045
10. Adams AS, Trinacty CM, Zhang F et al (2008) Medication adherence and racial differences in A1C control. Diabetes Care 31(5):916–921
11. Howard G, Lackland DT, Kleindorfer DO et al (2013) Racial differences in the impact of elevated systolic blood pressure on stroke risk. JAMA Intern Med 173(1):46–51
12. Hall EC, Segev DL, Engels EA (2013) Racial/ethnic differences in cancer risk after kidney transplantation. Am J Transplant 13(3):714–720
13. O'Campo P, Xue X, Wang MC et al (1997) Neighborhood risk factors for low birth weight in Baltimore: a multilevel analysis. Am J Public Health 87(7):1113–1118
14. Diez Roux AV, Merkin S, Arnett D et al (2001) Neighborhood of residence and incidence of coronary heart disease. N Engl J Med 345(2):99–106
15. Logan J (2002) Separate and unequal: the neighborhood gap for blacks and Hispanics in metropolitan America. Lewis Mumford Center for Comparative Urban and Regional Research, Albany
16. Krieger N, Chen J, Waterman P et al (2003) Race/ethnicity, gender, and monitoring socioeconomic gradients in health: a comparison of area-based socioeconomic measures – the public health disparities geocoding project. Am J Public Health 93(10):1655–1671

17. Darden JT (2007) Changes in black residential segregation in metropolitan areas of Michigan, 1990–2000. In: Darden JT, Stokes C, Thomas R (eds) The state of black Michigan, 1967–2007. Michigan State University Press, East Lansing, pp 147–160
18. Darden J, Rahbar M, Jezierski L et al (2010) The measurement of neighborhood socioeconomic characteristics and black and white residential segregation in Metropolitan Detroit: implications for the study of social disparities in health. Ann Assoc Am Geogr 100(1):137–158. doi:10.1080/00045600903379042
19. Jones K, Duncan C (1995) Individuals and their ecologies: analyzing the geography of chronic illness within a multilevel modeling framework. Health Place 1:7–30
20. Ellen IG, Turner MA (1997) Does neighborhood matter? Assessing recent evidence. Hous Policy Debate 8(4):833–866
21. Oakes JM (2004) The (mis)estimation of neighborhood effects: causal inference for a practicable, social epidemiology. Soc Sci Med 58:1929–1952
22. Katz LF, Kling JR, Leibman JB (2001) Moving to opportunity in Boston: early results of a randomized mobility experiment. Q J Econ 116(2):607–654
23. Kling J, Liebman J, Katz L et al (2004) Moving to opportunity and tranquility: neighborhood effects on adult economic self-sufficiency and health from a randomized housing voucher experiment, Faculty working paper series no. RWP04–035. John F. Kennedy School of Government, Cambridge, MA
24. Sampson RJ, Wilson WJ (1995) Toward a theory of race, crime and urban inequality. In: Hagan J, Peterson RD (eds) Crime and inequality. Stanford University Press, Stanford, pp 37–54
25. Lanctot N, Smith CA (2001) Sexual activity, pregnancy and deviance in a representative urban sample of African American girls. J Youth Adolesc 30:349–372
26. Sampson RJ, Raudenbush SW (1999) Systematic social observation of public spaces: a new look at disorder in urban neighborhoods. Am J Sociol 105:603–651
27. H¨agerstrand T (1970) What about people in regional science? Pap Reg Sci Assoc 24:7–21
28. Kwan M-P (1999) Gender and individual access to urban opportunities: a study using space–time measures. Prof Geogr 51:211–227
29. Stafford M, Martikainen P, Lahelma E et al (2004) Neighbourhoods and self-rated health: a comparison of public sector employees in London and Helsinki. J Epidemiol Community Health 58:772–778
30. Ross N, Wolfson M, Dunn J et al (2000) Relations between income inequality and mortality in Canada and the United States: cross sectional assessment using census data and vital statistics. Br Med J 320:898–902
31. Galea S, Ahern J, Rudenstine S et al (2005) Urban built environment and depression: a multilevel analysis. J Epidemiol Community Health 59(10):822–827
32. Houston D, Wu J, Ang P, Winer A (2004) Structural disparities of urban traffic in Southern California: implications for vehicle-related air pollution exposure in minority and high poverty neighborhoods. J Urban Aff 26(5):565–592
33. Savitz DA, Feingold L (1989) Association of childhood cancer with residential traffic density. Scand J Work Environ Health 15:360–363
34. Oosterlee A, Drijver M, Lebret E et al (1996) Chronic respiratory symptoms of children and adults living along streets with high traffic density. Occup Environ Med 53:243–257
35. Van Vliet P, Knape M, de Hartog J et al (1997) Motor vehicle exhaust and chronic respiratory symptoms in children living near freeways. Environ Res 74(2):122–132
36. Wilhelm M, Ritz B (2003) Residential proximity to traffic and adverse birth outcomes in Los Angeles County, California 1994–1996. Environ Health Perspect 111:207–216
37. May JM (1950) Medical geography: its methods and objectives. Geogr Rev 40:9–41
38. May JM (1958) The ecology of human disease: studies in medical geography. MD Publications, New York
39. Audy JR (1958) Medical ecology in relation to geography. Br J 12:102–110
40. Barrett FA (2002) The role of French-language contributors to the development of medical geography (1782–1933). Soc Sci Med 55:155–165

41. Dubos RJ (1987) Mirage of health: utopias, progress and biological change. Rutgers University Press, New Brunswick
42. Hunter JM (1974) The challenge of medical geography. In: Hunter JM (ed) The geography of health & disease. Department of Geography, University of North Carolina, Chapel Hill, pp 1–31
43. Meade MS, Emch M (2010) Medical geography. Guilford, New York
44. Mayer JD (2000) Geography, ecology and emerging infectious diseases. Soc Sci Med 50:937–952
45. Dubos RJ (1965) Man adapting. Yale University Press, New Haven
46. Carrel M, Emch M (2013) Genetics: a new landscape for medical geography. Ann Assoc Am Geogr. doi:10.1080/00045608.2013.784102
47. Scott CA, Robbins PG, Comrie AC (2012) The mutual conditioning of humans and pathogens: implications for integrative geographical scholarship. Ann Assoc Am Geogr 102(5):977–985. doi:10.1080/00045608.2012.657511
48. Aitken S (2001) Shared lives: interviewing couples, playing with their children. In: Limb M, Dwyer C (eds) Qualitative methodologies for geographers: issues and debates. Arnold, London, pp 73–86
49. Kobayashi A (2001) Negotiating the personal and the political in critical qualitative research. In: Limb M, Dwyer C (eds) Qualitative methodologies for geographers: issues and debates. Arnold, London, pp 55–70
50. Valentine G (2002) People like us: negotiating sameness and difference in the research process. In: Flowerdew R, Martin D (eds) Methods in human geography. Longman, Harlow, pp 116–126
51. Dwyer C, Limb M (2001) Introduction: doing qualitative research in geography. In: Limb M, Dwyer C (eds) Qualitative methodologies for geographers: issues and debates. Arnold, London, pp 1–20
52. Winchester HPM (2005) Qualitative research and its place in human geography. In: Hay I (ed) Qualitative research methods in human geography. Oxford University Press, South Melbourne, pp 3–18
53. Cope M (1998) Home–work links, labor markets, and the construction of place in Lawrence, Massachusetts, 1920–1939. Prof Geogr 50(1):126–140
54. Sparke M (1998) Mapped bodies and disembodied maps: (dis)placing cartographic struggle in colonial Canada. In: Nast HJ, Pile S (eds) Places through the body. Routledge, New York, pp 305–337
55. Creswell JW (2007) Qualitative inquiry and research design: choosing among five approaches. Sage, London
56. Bell S, Reed M (2004) Adapting to the machine: integrating GIS into qualitative research. Cartographica 39(1):55–66
57. Kwan M-P, Knigge L (2006) Doing qualitative research using GIS: an oxymoronic endeavor? Environ Plan A 38(11):1999–2002
58. Pavlovskaya ME (2006) Theorizing with GIS: a tool for critical geographies? Environ Plan A 38(11):2003–2020
59. Mugerauer R (2000) Qualitative GIS: to mediate, not dominate. In: Janelle DG, Hodge DC (eds) Information, place, and cyberspace: issues in accessibility. Springer, Berlin, pp 317–325
60. Kwan M-P, Ding G (2008) Geo-narrative: extending geographic information systems for narrative analysis in qualitative and mixed-method research. Prof Geogr 60:443–465. doi:10.1080/00330120802211752
61. Matthews SA, Detwiler J, Burton L (2005) Geo-ethnography: coupling geographic information analysis techniques and ethnographic methods in urban research. Cartographica 40(4):75–90
62. Knigge L, Cope M (2006) Grounded visualization: integrating the analysis of qualitative data through grounded theory and visualization. Environ Plan A 38(11):2021–2037
63. Boschmann EE, Cubbon E (2013) Sketch maps and qualitative GIS: using cartographies of individual spatial narratives in geographic research. Prof Geogr. doi:10.1080/00330124.2013.781490

64. Valentine G, Sadgrove J (2012) Lived difference: a narrative account of spatiotemporal processes of social differentiation. Environ Plan A 44(9):2049–2063
65. Ahas R, Silm S, Järv O et al (2010) Using mobile positioning data to model locations meaningful to users of mobile phones. J Urban Technol 17:3–27
66. Almanza E, Jerrett M, Dunton G et al (2012) A study of community design, greenness, and physical activity in children using satellite, GPS and accelerometer data. Health Place 18:46–54
67. Richardson DB, Volkow ND, Kwan M-P et al (2013) Spatial turn in health research. Science 339(6126):1390–1392
68. Wiehe SE, Kwan M-P, Wilson J et al (2013) Adolescent health-risk behavior and community disorder. PLoS ONE 8(11):e77667. doi:10.1371/journal.pone.0077667

Chapter 5
Data Privacy and Ethical Uses of Volunteered Geographic Information

Abstract There is great potential for volunteered geographic information (VGI) to augment data used for public health disease surveillance, in areas such as mass gatherings and qualitative GIS. The goal of this chapter is to explore these important issues of patient privacy, ethics, and liability, as they pertain to the use of VGI to augment health information exchanges (HIEs) in providing data for public health research programs. The current attention on health reform and HIEs provide professional geographers with an excellent opportunity to explore the contributions of VGI to this field. The chapter begins by briefly describing the legislation of patient privacy and protection in the United States, such as the Health Insurance Portability and Accountability Act of 1996 (HIPAA) and the American Recovery and Reinvestment Act of 2009. It also discuss the appropriate and inappropriate disclosures of protected health information (PHI). Next, it examines the ethical and legal issues surrounding the use of VGI in disease surveillance. Finally, the chapter will demonstrate that VGI yields tremendous value in providing sensitive and timely surveillance data when reliable and consistent communications between health care providers and regional health authorities are not possible.

Keywords Volunteered geographic information • Protected health information • Ethics • Data liability • HIPAA • Patient privacy

5.1 Introduction

Volunteered geographic information (VGI) a term is attributed to Dr. Michael Goodchild, a professional geographer who coined it in 2007 to describe the practice of observing, collecting and producing geographic information by people who have no formal training in geospatial data [1]. VGI has gained prominence in the general public because of Web 2.0 – i.e., the online technologies that have transformed the Internet into a platform for the collaboration, production and distribution of information. Examples of Web 2.0 technologies include geospatial browsers, mobile devices, and cloud computing services. Cloud computing services currently allow users to conveniently access a shared pool of computing resources – e.g., networks, servers, storage, and applications – that can be rapidly deployed with minimal

A.J. Blatt, *Health, Science, and Place*, Geotechnologies and the Environment 12,
DOI 10.1007/978-3-319-12003-4_5

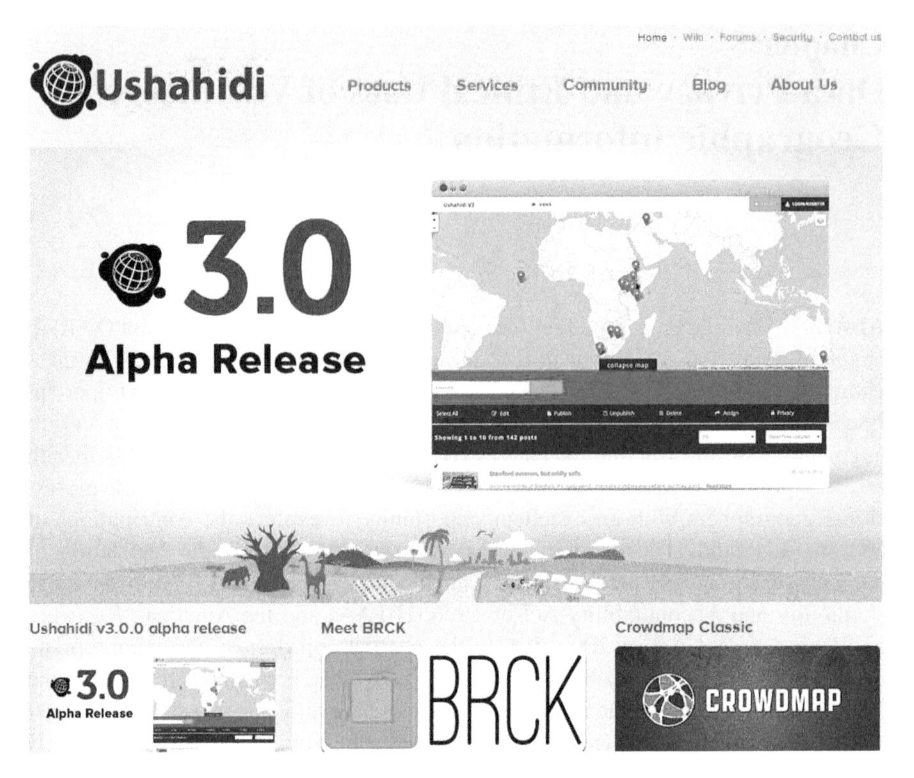

Fig. 5.1 Ushahidi is a non-profit technology company that demonstrates the use of VGI in crisis management (http://www.ushahidi.com/, Accessed 17 Nov 2014)

effort or technical support. When used appropriately, Web 2.0 transforms data consumers into data producers (i.e., map-makers and cartographers) [2].

In recent years, VGI has been used to map crisis situations, such as the Haitian earthquakes in 2010, and relief efforts in forest fires, floods, hurricanes and other natural disasters (Fig. 5.1). After the volunteer contributors capture and record information using their own mobile devices (such as smartphones and GPS units), they can submit their data to a website or central agency that analyzes, reviews and publishes the information on base maps provided by OpenStreetMap or Google Maps. While some authorities may question the reliability and accuracy of VGI-based maps, the timely production of near instantaneous geospatial information during times of crisis has proven to be invaluable [3, 4].

As with any new introduction of ideas, many legal and ethical issues surround this exchange of geospatial information, especially in the field of disease surveillance. One important issue concerning patient privacy involves the location-based devices that record the identities, positions, movements, and other characteristics of the contributors of the health information [5]. Another issue is the nature of these contributors of health information and the types of unstated agreements they may be entering into, by their volunteerism [6, 7]. Finally, public health officials and health care organizations are understandably concerned with the credibility of the health geographic data offered and the algorithms used to validate the truthfulness of VGI [8, 9].

The goal of this chapter is to explore these important issues of patient privacy, ethics, and liability, as there is a great potential for VGI to augment health information exchanges (HIEs) in providing data for public health research programs. The attention on health reform and HIEs provide professional geographers with an excellent opportunity to explore the contributions of VGI to this field. To address these contributions, this chapter will briefly describe the legislation of patient privacy and protection in the United States. Then, it will present a candid discussion of the ethical and legal issues surrounding VGI in disease surveillance, and conclude with a discussion of the approaches and solutions that only an engaged community of professional geographers can propose.

5.2 HIPAA Legislation and Patient Privacy Protections

There have been several significant legislative actions to protect individual privacy, as it pertains to HIEs. First, the Privacy Rule in the Health Insurance Portability and Accountability Act of 1996 (HIPAA) created new procedural requirements for the use of protected health information (PHI) and defined PHI more broadly than had been done in the past [10, 11]. Second, the American Recovery and Reinvestment Act of 2009 directed the new National Coordinator for Health Information Technology to pay closer attention to both data access and data confidentiality matters, as they relate to the HIEs [12]. Finally, the Patient Protection and Affordable Care Act of 2010 directed the same office to "develop national standards for the management of data collected," and "develop interoperability and *security systems* for data management" [13].

Among these, the one that bears the greatest impact on patient privacy protection is HIPAA's Privacy Rule, which went into full effect in 2003. The Privacy Rule defines and protects PHI, which is any information that relates to the individual's past, present, or future physical or mental health or condition, the provision of healthcare to that individual, or the payment for the provision of healthcare to the individual. Simply put, this includes any information that identifies an individual or for which there is a reasonable basis to believe it can be used to identify an individual. A list of common identifiers that are considered PHI is listed in Table 5.1.

While permissible uses and disclosures of PHI include those made for purposes of treatment, payment and healthcare operations, reasonable care must be taken to avoid disclosures. The U. S. Department of Health and Human Services' Office of Civil Rights (OCR) is responsible for enforcing HIPAA. OCR's investigations are complaint driven. In response to complaints received, OCR conducts an investigation and may require corrective action and/or a resolution agreement, if a HIPAA violation has occurred. OCR may impose civil monetary penalties and refer matters to the US Department of Justice for criminal investigations.

Table 5.1 18 identifiers of protected health information, or PHI, in accordance with 45 CFR 164.514[a]

1. Names
2. Geographic subdivisions smaller than a state (except the first three digits of a zip code if the geographic unit formed by combining all zip codes with the same three initial digits contains more than 20,000 people and the initial three digits of a zip code for all such geographic units containing 20,000 or fewer people is changed to 000)
3. All elements of dates (except year) for dates directly related to an individual, including birth date, admission date, discharge date, and date of death and all ages over 89 and all elements of dates (including year) indicative of such age (except that such ages and elements may be aggregated into a single category of age 90 or older)
4. Telephone numbers
5. Fax numbers
6. Electronic mail addresses
7. Social security numbers
8. Medical record numbers
9. Health plan beneficiary numbers
10. Account numbers
11. Certificate/license numbers
12. Vehicle identifiers and serial numbers, including license plate numbers
13. Device identifiers and serial numbers
14. Web Universal Resource Locators (URLs)
15. Internet Protocol (IP) address numbers
16. Biometric identifiers, including finger and voice prints
17. Full face photographic images and any comparable images
18. Any other unique identifying number, characteristic, or code (excluding a random identifier code for the subject that is not related to or derived from any existing identifier)

[a]Please note that de-identified health information entails the removal of all of the above identifiers as well as the absence of actual knowledge by the covered entity that the remaining information could be used alone or in combination with other information to identify the individual

In January 2013, HIPAA was updated to include the Security Rule and Breach Notification portions of the Health Information Technology for Economic and Clinical Health (HITECH) Act. In these updates, the greatest changes relate to the expansion of requirements to include business associates, as opposed to the covered entities that were originally mentioned in the law. Additionally, the definition of "significant harm" to an individual in the analysis of a breach was updated to provide increased scrutiny to covered entities with the intent of disclosing more breaches which had been previously gone unreported. As an example, an organization previously needed to demonstrate that harm had occurred to prove a security breach. Now, they must prove that harm had not occurred. Other changes include the protection of PHI for 50 years after death, and the imposition of more severe penalties for PHI violations [14].

In addition, independent institutional review boards (IRBs) have been created to protect patients' rights and risk from physical or psychological harm in research. IRBs are independent ethics review committees which are formally designated to approve, monitor, and review biomedical and behavioral research involving humans. They often perform some type of risk-benefit analysis to determine whether or not the proposed research should be conducted [15]. In the United States, regulations from the Food and Drug Administration (FDA) and Department of Health and Human Services (Office for Human Research Protections) have empowered IRBs to approve, require modifications in planned research prior to approval, or disapprove research. IRBs are responsible for critical oversight functions for research conducted on human subjects that are "scientific," "ethical," and "regulatory" [16].

5.3 Legal and Ethical Issues Concerning Volunteered Geographic Information

According Dr. Francis Harvey, a noted critical geographer, VGI can be distinguished from contributed geographic information (CGI) by determining whether the collected geographic information was provided from a data producer who opted "in" to contribute the data. For instance, geographic information is "volunteered," if a geotagged photograph or observational note was actively and willingly uploaded onto OpenStreetMap (Fig. 5.2). On the other hand, geographic information is "contributed" if the data producer had an opportunity to opt "out" of the data production process (as in a remotely sensing image used for surveillance) [17]. This difference speaks to the importance of distinguishing crowdsourced information collected by people actively deciding to collect data (and having some measure of control), from crowdsourced data collected automatically or with a clear abrogation of possibilities to influence the collection and reuse of the data.

Regardless of whether the geographic information was volunteered or contributed, these issues of liability, privacy, and ethics are very relevant to disease surveillance, especially when untrained citizens are contributing data to inform public health. One of the most prominent uses of VGI in disease surveillance is Google Flu Trends, which was developed in 2008 by Google and the Centers for Disease Control and Prevention (CDC) [18]. Google Flu Trends relies on the data mining records of flu-related search terms entered in Google's Internet search engine, and sophisticated computer simulations. Since its inception, estimates from Google Flu Trends have closely matched the CDC's own surveillance data over time, often delivering them several days faster than the CDC surveillance systems can. The success of Google Flu Trends has been so great that it has been rolled out to 29 countries and extended to include surveillance for a second disease, dengue (Fig. 5.3).

In recent years, however, Google Flu Trends has missed its intended forecasts. Researchers speculate that these missed estimates may be due to widespread media coverage of a particularly severe flu season in one year, such as a declaration of a public-health emergency by New York State. The proliferation of press reports on the Internet may also have triggered many flu-related searches by people who were not ill [19].

Fig. 5.2 OpenStreetMap view of London (http://www.openstreetmap.org/note/15014, Accessed 17 Nov 2014)

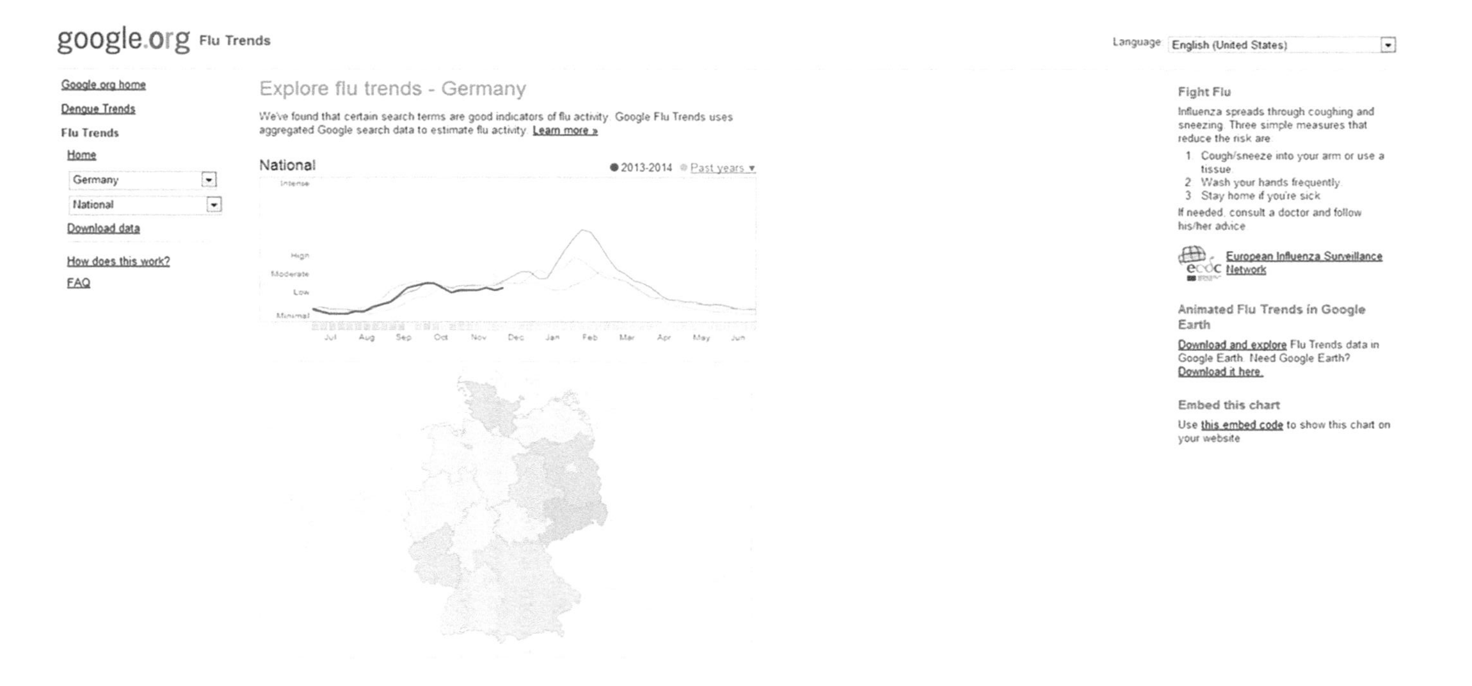

Fig. 5.3 Google Flu Trends in Germany (http://www.google.org/flutrends/de/#DE, Accessed 17 Nov 2013)

Liability issues arising from data accuracy are also very important, especially when there is a potential endangerment of life, as seen in the use of VGI in disease surveillance applications [20, 21]. Since the contributors of VGI often do not have any formal training or expertise in mapping or geography, mistakes and errors may exist in VGI-based disease surveillance services. For this reason, the contractual relationship between the VGI website and its users must be posted in the "Terms of Use" on the web page. For contributors, particular care must be exercised by not promising any type of warranty for the data submitted. Doing so may expose and heighten the risk of legal responsibility for any errors (intended or un-intended). Also, contributors should not provide any indemnification for the data and information submitted. This will avoid having to defend indemnity in the courts on behalf of the VGI website.

Furthermore, VGI service providers need to ensure that they do not engage in negligent conduct and implement a secure process to validate volunteer contributions. Because it can be difficult to determine whether a VGI website is a product, service, or mere information, a VGI service may be liable in cases where accuracy was promised and was breached, resulting in a causal link between VGI provider and the loss suffered [22]. The exercise of due diligence from a reasonable person is the common threshold test for this type of liability.

In addition, there are many ethical and privacy issues involved with data mining public records from the Internet for disease surveillance research. Some people participating in health-related discussion forums have no problems with the constant and uncontrolled collection of their personal information, while others feel that this intrusion is similar to situations where governments have secretly collected information about citizens [23, 24]. Unfortunately, research using VGI to uncover health trends in public discussion forums are exempt from IRB applications because "research involving the collection or study of existing data, documents, records, pathological specimens, or diagnostic specimens, if these sources are publicly available" are exempt from IRB review [15]. As a result, personal health information that is volunteered and is geo-tagged by an Internet Protocol (IP) address is not protected by the independent ethics review committees that regularly review research protocols involving human subjects.

Furthermore, as crowdsourced geographic information becomes more prevalent in certain parts of our society, more detailed data is constantly being collected from smart phone users, unless the phones are turned off or the location services are disabled. There are many examples of how crowdsourced locational data is collected, often without the data producers' knowledge and frequently without information about its reuses [25–27]. For instance, according to a CNET news report, Microsoft has collected spatial data on laptops, cell phones, and Wi-Fi devices and released that information on the Internet [28]. In public health applications, the practice of removing attribute data from a person's geographic footprint and thereby only linking things by geographic and temporal proximity should not be interpreted as an adequate mode of protecting an individual's privacy. Reverse geocoding, a process of identifying a street address from a point on a map, allows for the identification of an individual, as do time-stamped geographic data or travel paths [29–31]. In addition, methods have been developed to reverse identify the addresses of infected patients via low-resolution images (e.g., 50 dots-per-inch and 1:100,000 scale) published in medical journals [32]. Hence, crowdsourced geographic information plays an important role in the ability of companies and government agencies to know and predict people's activities, and – to a certain extent – their health conditions.

5.4 Collaborating Towards a Solution: New Use Cases for Volunteer Geographic Information

At the beginning of this chapter, three important issues were raised concerning the privacy, accuracy, and use of VGI in scientific research. First, given the nature of VGI (and CGI), how can personal privacy be safeguarded in a forum that is publicly available, such as the Internet? Second, how can VGI be verified, as to its truth and accuracy, especially in the context of public health disease surveillance? Finally, how can VGI be used to supplement traditional sources of health information, especially in the context of HIEs?

These issues are most adequately addressed by collaborations across disciplines and governing bodies. Several solutions are available in the broader realm of public health and epidemiological research; however, they need to be adjusted to address the special nature of geospatial data. Addressing personal privacy on the Internet is a difficult issue because, while it is true that the Internet is a public forum and there is no true protection of privacy once VGI is posted there, a VGI-based disease surveillance service could choose to exercise caution and responsibility when collecting this information and separate the PHI from the medical records as the source of data collection and perform the analysis on the de-identified records. This protocol has been suggested for traditional, notification-based surveillance systems and cancer registries, and the current generation of information technology does easily support this type of architecture [33].

Several algorithms have been developed to verifying the truthfulness and data accuracy of VGI-based disease surveillance services. There are two fundamental approaches to assessing the reliability of VGI. The first approach is the simplest, whereby all VGI is assumed to be either entirely credible or not until it is confirmed or rejected by a subsequent report. In this approach, all data producers are treated equally with respect to their prior knowledge/skills; and their reliability is assessed by their peers through the creation of informal social networks of influence and trust [34, 35]. The second approach considers the skill set and prior credibility of the data producer offering the VGI, and produces a rating or score of each contributor based on these factors. In this approach, the reliability of the VGI is dependent on the prior assessment of the data producer and previous VGI offered by this individual to the system. If a number of prior contributions have been rejected, then the individual tagged with a low reliability score that may result in the automatic rejection of any subsequent contributions made (unless the contributions can be independently confirmed by a third party in the future). On the other hand, if the data producer has a history of high quality submissions that are consistently confirmed, then he or she may be given a high credibility score, leading to an automatic acceptance of the VGI into the database [36–38].

In addition, as Dr. Francis Harvey has suggested, truth in labeling can be very helpful, as it relates to the provenance of VGI, in recording and making accessible characteristics of the collection, processing, and reuses of information to existing and potential data consumers. As part of provenance, drawing a distinction between CGI and VGI can be an important truth-in-labeling indicator for potential users about quality and potential biases [17].

One of the greatest potential applications of VGI is to provide sensitive and timely surveillance data when reliable and consistent communications between health care providers and regional health authorities are not possible. VGI presents a viable approach to collect disease incidence data – or other data (e.g., climatic and environmental risks) that directly influences the prevalence of disease. VGI uses the local citizenry to collect surveillance information, thereby increasing the number of resources available to collect the data. If deployed correctly, it may not be necessary to dispatch health professionals to obtain the data directly, as in mass gathering.

References

1. Goodchild MF (2007) Citizens as voluntary sensors: spatial data infrastructure in the world of Web 2.0. Int J Spat Data Infrastruct Res 2:24–32
2. Sui D, Elwood S, Goodchild MF (2013) Crowdsourcing geographic knowledge: volunteered geographic information in theory and practice. Springer, Berlin
3. Goodchild MF, Glennon JA (2010) Crowdsourcing geographic information for disaster response: a research frontier. Int J Digit Earth 3(3):231–241
4. Levental S (2012) A new geospatial services framework: how disaster preparedness efforts should integrate neogeography. J Map Geogr Lib 8(2):134–162
5. The Economist (2010) Dicing with data. The Economist online, 20 May 2010. Available at http://www.economist.com/node/16163396. Accessed 10 Nov 2013
6. Dunn C (2007) Participatory GIS – a people's GIS? Prog Hum Geogr 31(5):616–637
7. McDougall K (2009) Volunteered geographic information for building SI. In: Ostendorf B, Baldock P, Bruce D, Burdett M, Corcoran P (eds) Proceedings of the Surveying and Spatial Sciences Institute: biennial international conference, Adelaide, Australia 2009. Surveying and Spatial Sciences Institute, Hobart, pp 645–653
8. Flanagin AJ, Metzger MJ (2008) The credibility of volunteered geographic information. GeoJournal 72:137–148
9. Lane J, Schur C (2010) Balancing access to health data and privacy: a review of the issues and approaches for the future. Health Serv Res 45(5 part 2):1456–1567
10. Health Insurance Portability and Accountability Act of 1996. Pub. L. 104–191 (Aug. 21, 1996).
11. American Statistical Association Open Letter on Proposed Changes to CMS Part D Public Use Files, 17 September 2008. Available at http://www.amstat.org/outreach/pdfs/CMSPartDPUF.pdf. Accessed 17 Nov 2013
12. Public Law 111–5 (2009) Available at http://www.gpo.gov/fdsys/pkg/PLAW-111publ5/pdf/PLAW-111publ5.pdf. Accessed 17 Nov 2013
13. PPACA Title XXXI, Section 3101
14. HITECH Act, 45 CFR Parts 160 and 164

15. Price PL (2012) Geography, me, and the IRB: from roadblock to resource. Prof Geogr 64(1):34–42
16. National Research Act of 1974, Title 45 CFR Part 46
17. Harvey F (2013) To volunteer or to contribute locational information? Towards truth in labeling for crowdsourced geographic information. In: Sui D, Elwood S, Goodchild MF (eds) Crowdsourcing geographic knowledge: volunteered geographic information in theory and practice. Springer, Berlin, pp 31–42
18. Google.org Flu Trends. Available at http://www.google.org/flutrends/about/how.html. Accessed 17 Nov 2013
19. Butler D (2013) When Google got flu wrong. Nat New 494:155–156
20. Scassa T (2013) Legal issues with volunteered geographic information. Can Geogr 57(1):1–10
21. Mooney P, Corcoran P, Ciepluch B (2013) The potential for using volunteered geographic information in pervasive health computing applications. J Ambient Intell Humaniz Comput 4(6):731–745
22. Chandler JA, Levitt K (2011) Spatial data quality: the duty to warn users of risks associated using spatial data. Alberta Law Rev 49(1):79–106
23. Pew Research Center (2010) The future of online socializing. Available at http://pewresearch.org/pubs/1652/social-relations-online-experts-predict-future. Accessed 17 Nov 2013
24. Kar B, Crowsey RC, Zale JJ (2013) The myth of location privacy in the United States: surveyed attitude versus current practices. Prof Geogr 65(1):47–64
25. Acohido B (2011) Privacy implications of ubiquitous digital sensors. USA Today, 26 January 2011, P1B
26. Liptak A (2011) Court case asks if 'Big Brother' is spelled GPS. The New York Times, online. Available at http://www.nytimes.com/2011/09/11/us/11gps.html. Accessed 17 Nov 2013
27. National Research Council. (2007) Putting people on the map: protecting confidentiality with linked social-spatial data. National Academy Press, Washington, DC
28. McCullagh D (2011) Microsoft's web map exposes phone, PC locations. CNET. Available at http://news.cnet.com/8301-31921_3-20085028-281/microsofts-web-map-exposes-phone-pc-locations/. Accessed 17 Nov 2013
29. Brownstein J, Cassa C, Kohane I et al (2005) Reverse geocoding: concerns about patient confidentiality in the display of geospatial health data. In: AMIA annual symposium proceedings, Washington, DC, p 905
30. El Emam K, Brown A, AbdelMalik A et al (2010) A method for managing re-identification risk from small geographic areas in Canada. BMC Med Inf Decis Mak 10:18
31. Chow TE, Lin Y, Chan WD (2011) The development of a web-based demographic data extraction tool for population monitoring. Trans GIS 15(4):479–494
32. Brownstein JS, Cassa CA, Kohane IS et al (2006) An unsupervised classification method for inferring original case locations from low-resolution disease maps. Int J Health Geogr 5:56. doi:10.1186/1476-072X-5-56
33. Churches T (2003) A proposed architecture and method of operation for improving the protection of privacy and confidentiality in disease registers. BMC Med Res Methodol 3:1
34. Bishr M, Kuhn W (2007) Geospatial information bottom-up: a matter of trust and semantics. In: Fabrikant SI, Wachowicz M (eds) Lecture notes in geoinformation and cartography. Springer, Berlin, pp 365–387. doi:10.1007/978-3-540-72385-1_22
35. Bishr M, Mantelas L (2008) A trust and reputation model for filtering and classifying knowledge about urban growth. GeoJournal 72(3):229–237. doi:10.1007/s10708-008-9182-4
36. Crosetto M (2001) Uncertainty and sensitivity analysis: tools for GIS-based model implementation. Int J Geogr Inf Sci 15(5):415–437. doi:10.1080/13658810110053125
37. Conati C (2004) How to evaluate models of user affect? In: André E, Dybkjær L, Minker W, Heisterkamp P (eds) Affective dialogue systems, vol 3068. Springer, Berlin, pp 288–300. doi:10.1007/978-3-540-24842- 2_30
38. Langley S, Messina J, Grady S (2013) Utilizing volunteered information for infectious disease surveillance. Int J Appl Geospatial Res 4(2):54–70

Part III
Geospatial Data and Technologies

Chapter 6
Collaborative Mapping

Abstract This chapter introduces the role of Web 2.0, Web 3.0, and data mashups in use cases of collaborative mapping in public health disease surveillance. Current advances in these geospatial technologies enable geographers – both professional and citizen geographers – to contribute, analyze and map data from anywhere in the world. The role of "public participation GIS" and "participatory GIS" are examined in the context of volunteered geographic information (VGI), and their application in understanding the role of place in affecting human health. This integration of spatial data, collaboration, and decision-making is known as collaborative spatial decision-making (CSDM). There are a number of challenges to incorporating collaborative tools in spatial decision-making – e.g., a lack of interoperability with other tools commonly used by organizations, and a lack of generality to enable sustainable processes within a group accomplishing a specific task. A Spatial Data Infrastructure (SDI) presents a possible solution to these problems, as it promotes data sharing through data harmonization and standardization (thus avoiding unnecessary geospatial data duplication), at different organizational and geographical levels. Several organizations, such as the Association of American Geographers (AAG) and the National Institutes of Health (NIH), are collaborating to address the role of spatial and spatiotemporal data analyses and geographic computing resources in health and science research.

Keywords Public participation GIS • Participatory GIS • Web 2.0 • Web 3.0 • Data mashups • Spatial data infrastructures

6.1 Introduction

Understanding the role of place in affecting human health is an important aspect of understanding and preventing the spread of disease. Social and environmental factors contributing to the emergence and spread of disease include climate, water quality, employment and education settings, pollution, and a host of other human-environment interactions. The collection and management of vast amounts of geospatial data can be arduous and labor-intensive; however, the current advances in collaborative mapping (e.g., mashups and Web 2.0/3.0) provide innovative tools and opportunities that make it easier and more intuitive for geographers to contribute,

A.J. Blatt, *Health, Science, and Place*, Geotechnologies and the Environment 12,
DOI 10.1007/978-3-319-12003-4_6

analyze, and map data from anywhere in the world [1]. For a list of definitions and examples of terms that are commonly used in discussing collaborative innovations, please see Table 6.1.

Table 6.1 Terms used to describe activities related to collaborative mapping, their definitions, and examples

1.	AJAX: Stands for "Asynchronous JavaScript and XML." AJAX is a group of inter-related Web development tools used for creating interactive Web applications. A primary characteristic is the enhanced responsiveness and interactivity of Web pages which is achieved by exchanging small amounts of data with the host server (i.e., "behind the scenes") so that entire Web pages do not have to be reloaded each time there is a need to obtain or refresh data from the server. This feature increases the Web page's interactivity, speed, functionality, and usability.
	Example: Google Maps (http://maps.google.com/) is an example of AJAX application.
2.	Blog: A "Web journal" or "Web log." A blog is a specialized Web service that allows an individual (or group of individuals) to share a running log of events and personal insights with online audiences.
	Example: GIS and Science Blog (http://gisandscience.com/) attempts to record projects and news stories that demonstrate the applications of geospatial technology for scientific research and understanding (Fig. 6.1).
3.	Folksonomy: Also known as "social tagging." Folksonomy is the practice of collaboratively creating and managing tags to annotate and categorize content. In contrast to traditional subject indexing, metadata is not only generated by trained experts but also by the creators and consumers of the content.
	Example: Del.icio.us (http://del.icio.us/) is a social bookmarking Web service for storing, sharing and discovering Web bookmarks. Users can tag each of their bookmarks with a number of freely chosen keywords.
4.	Gadget/Widget: A mini-Web application on a Web page, blog or social profile that can quickly and easily provide visitors with user specific information and extra functionality. A gadget can be considered as a primitive widget.
	Example: Google Gadgets (e.g., calculator, calendar and thermometer) are miniature objects that offer dynamic content that can be placed on a Web page.
5.	Information commons: Provides access to information resources by a community of producers and consumers in an open access environment.
	Example: Pathway Commons (http://www.pathwaycommons.org/) serves as a central point of access to biological pathway information collected from public pathway databases.
6.	Mashup (in the Web context): A Web application that combines data and/or functionality from more than one source.
	Example: Geocommons (http://www.geocommons.com/) provides geo-mashup by providing a Web interface that allows users to select different maps and overlay them one on top of the other (Fig. 6.2).
7.	Ontology: A representation of concepts with a domain and the relationships between those concepts. It is a shared conceptualization of a domain.
	Example: Gene ontology (http://www.geneontology.org) is a popularly used ontology in biomedical informatics. It provides a controlled vocabulary to describe gene and gene product attributes in any organism. It involves three categories of information, namely, biological processes, molecular functions, and cellular locations.
8.	Social networking: A phenomenon defined by linking people to each other in some way. Users may work together to rate news and are linked by rating choices or explicit identification of other members.
	Example: Facebook (http://www.facebook.com/) is an example of a social networking. It is a Web site for people to discover and share content from anywhere on the Web.

(continued)

Table 6.1 (continued)

9.	Tag cloud: A visual depiction of user-generated tags. The importance or popularity of a tag is shown with font size or color. For example, the bigger the font size, the more popular the tag (Fig. 6.3).
10.	Web 2.0: Includes the following key features: (1) user-centric and user-oriented; (2) Web services, Web APIs; (3) widgets, gadgets, mashups; (4) blogs, feeds, Wiki's, tagging; (5) social networking; (6) client-rich technologies like AJAX.
	Examples: Web sites like Flickr (http://www.flickr.com/), YouTube (http://www.youtube.com/), and MySpace (http://www.myspace.com/) possess Web 2.0 features.
11.	Web 3.0: Used to describe the future of the World Wide Web. Following the introduction of the phrase Web 2.0 as a description of the recent evolution of the Web, people have used the term Web 3.0 to hypothesize about a future wave of Internet innovation.
	Example: Semantic Web is a kind of Web 3.0 technology extending the Web such that the semantics of information and services on the Web is defined, making it possible for the Web to understand and satisfy the requests of people and machines using the Web content.
12.	Web service: A software system designed to support interoperable machine-to-machine interaction over a network. Web services are frequently just Web APIs that can be accessed over a network, such as the Internet, and executed on a remote system hosting the requested services.
	Example: SOAP (Simple Object Access Protocol) is an XML-based protocol for accessing Web services over HTTP (HyperText Transfer Protocol).
13.	Wiki: A wiki is a software program that allows users to create collaborative websites and share content about a particular project or topic.
	Example: Wikipedia (http://www.wikipedia.org/) facilitates the community-based creation and curation of knowledge.

Fig. 6.1 GIS and Science Blog, a Web log that collects information about projects which communicate the value of using GIS technology to advance scientific understanding (http://gisandscience.com/, Accessed 25 May 2014)

Fig. 6.2 GeoCommons is an online mapping application that easily imports a variety of data formats, including geospatial data, and quickly produces sharable maps (http://geocommons.com/, Accessed 18 May 2014)

Fig. 6.3 An example of a tag cloud with terms related to Web 2.0 (http://en.wikipedia.org/wiki/Tag_cloud, Accessed 25 May 2014)

Web 2.0 describes World Wide Web sites that use technology beyond the static pages of earlier Web sites. Web 2.0 allow users to be producers, as well as consumers, of information. By increasing what was already possible in "Web 1.0," Web 2.0 allows users to provide content and exercise some control over that content. Examples of Web 2.0 sites include social networking sites (like Facebook and Twitter), user-authored Web sites, self-publishing platforms, tagging, and social bookmarking. In short, Web 2.0 sites promote on-line collaborations and the sharing of information [2].

While Web 2.0 offers many user-friendly tools for contributing volunteered geographic information in public health applications (see Chap. 5), Web 3.0 improves the user experience by enabling computers to help users find and integrate information over the Internet in a more sophisticated way. This collection of third-generation, Internet-based services – such as the Semantic Web, natural language searches, data-mining, machine learning, recommendation agents, and artificial intelligence technologies – emphasize a machine-facilitated understanding of information so that the user can have a more productive and intuitive interactive experience with the information. Web 2.0 and Web 3.0 are not two conflicting visions; instead, they are complementary to each other [3, 4].

"Mashup" is a term originally used to refer to the mixing of musical tracks to create a new piece of music, such as combining the audio from one song with the vocal track from another. In Web development, a mashup is a Web page (or Web-enabled software application) that uses content from more than one source to create a single new service displayed in a single graphical interface. A simple example is combining the addresses and photographs of local library branches with a Google map to create a map mashup. The term "mashup" implies easy and fast integration, frequently using open application programming interfaces (open API) and data sources to produce enriched results that were not necessarily the original reason for producing the raw source data [5].

Mashups are becoming increasingly widespread, especially in the context of combining geographic data and displaying such integrated data on maps. Studies that involve the association between human diseases (e.g., cancer) and environmental factors often require the integration of disparate data sources such as population census, air quality and environmental pollution release, and health care utilization data. As these disparate data streams are typically produced by different agencies, there are a number of political and technological challenges to automating the integration of data from these sources. Web 2.0/3.0 mashups provide users with the tools to integrate disparate health care data, for enhancing environmental health research and disease surveillance (Fig. 6.2).

In a popular example known as "GeoCommons," the researchers identified a cancer profile dataset from the National Cancer Institute and a "heat" map that details the number of polluted rivers/streams in the United States (where a brighter color corresponds to a higher number of polluted rivers/streams). The GeoCommons interface enabled the state cancer profile map to be superimposed with the water pollution map, and illustrated that most of the states with high cancer death rates are in the fire zone [6].

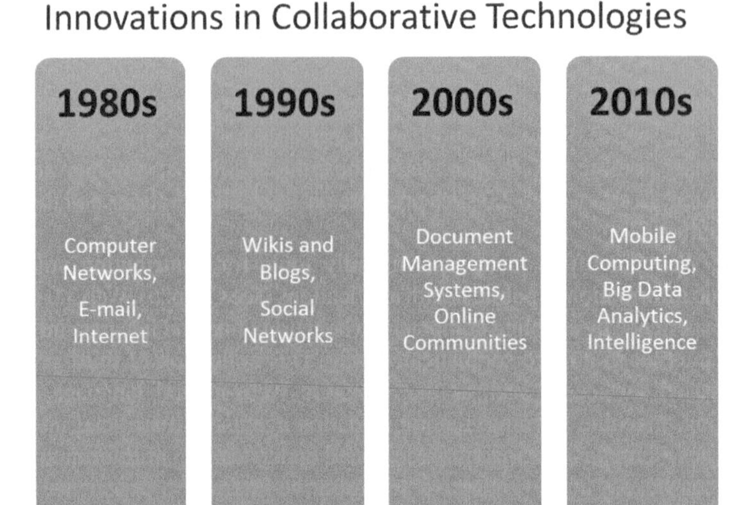

Fig. 6.4 The evolution of collaborative innovations

In disease surveillance, Web-based GIS technology provides quick access to distributed data for the analysis, visualization, planning, and modeling of the spread of disease. Since Web-based GIS promotes a near-real time response to pandemics and epidemics, it is effective for understanding disease phenomena to support decision making [7–9].

The goal of this chapter is to describe some of the current challenges that exist in these innovations in Web development (Fig. 6.4), as well as their current and potential future applications in public health. Towards this end, the chapter will discuss these topics against a backdrop of the current advances in geospatial data analytics and geovisualization.

6.2 Web 2.0, Web 3.0, and Mashups

The term "participatory GIS" (PGIS) was first widely introduced in the mid-1990s to emphasize a shift towards a critical evaluation of the uses of GIS in society [10]. The term "public participation GIS" is associated with the use of GIS to foster grassroots involvement in policy decision-making [11–13]. This integration of spatial data, collaboration and decision-making is relatively recent. It is often referred to as collaborative spatial decision-making (CSDM) [14, 15]. CSDM expands current uses of geographical information systems (GIS) by introducing new ways of collecting spatial data, relating spatial data with other types of data, and displaying and analyzing the collected data according with their geographical context [16–18]. CSDM builds upon GIS with decision-making models and processes, task coordination, collaboration, and new types of collaboration data such as talks, chats, discussions, negotiations, and brainstorms [19, 20].

Researchers have identified several key components of a successful CSDM, such as obtaining management support, the role of formalized project plans and project champions, a technology that is useful and easy to use, and the commitment and satisfaction of the end-users with the decision-making process and end results [21–23]. However, a number of issues have made it difficult to incorporate collaborative tools in spatial decision-making, such as the lack of interoperability with other tools commonly used by organizations, the lack of generality to enable sustainable processes within a group accomplishing a specific task, and the design problems from specifications that are not clearly linked to the decision problems these systems are expected to support [24–26].

Web 2.0 provides an excellent foundation to implement the user-friendly online tools used in CSDM. The ease of use of Web 2.0 APIs stems from a combination of JavaScript for the actual programming functionality, and XML and the JavaScript Object Notation (JSON) as the preferred formats for data transfer. This combination, known as AJAX (Asynchronous JavaScript and XML), enables Web applications to behave more like desktop programs. End-users are now able to receive updated content from a Web site without having to wait for a reload of the entire page – overcoming a usability issue that was apparent in Web 1.0. AJAX-based user interfaces allow for seamless interaction with online applications. Publicly available geospatial data enable users to build PPGIS with simplified or no licensing requirements. In contrast to the complex and expensive Web service architectures that were established for business and commercial applications, Web 2.0 services rely on lightweight application programming interfaces (APIs). APIs are easy-to-use and made publicly accessible for no cost, thus allowing programmers to combine multiple services into a single mashup [27, 28] (Figs. 6.5 and 6.6).

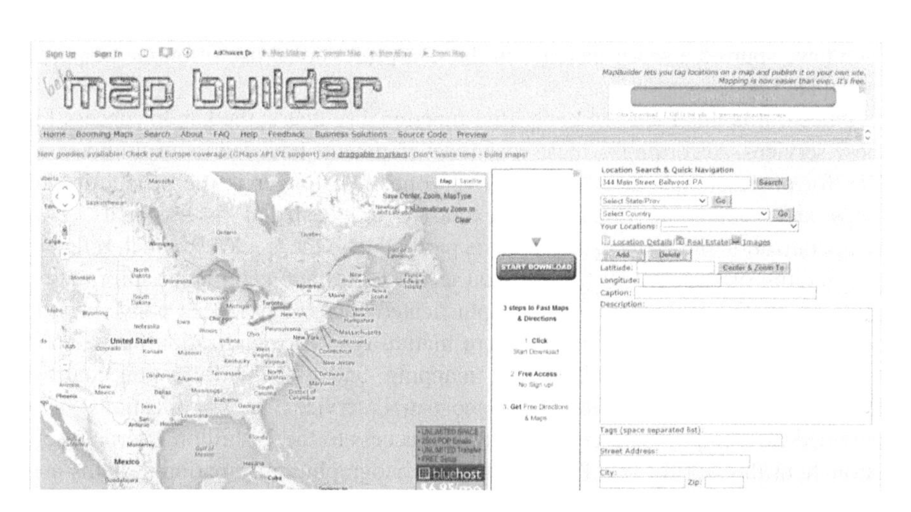

Fig. 6.5 MapBuilder, a Web2.0 service or rapid mashup development tool to build custom Google and Yahoo maps without any knowledge of the Google/Yahoo Maps API and JavaScript (http://www.mapbuilder.net/, Accessed 25 May 2014)

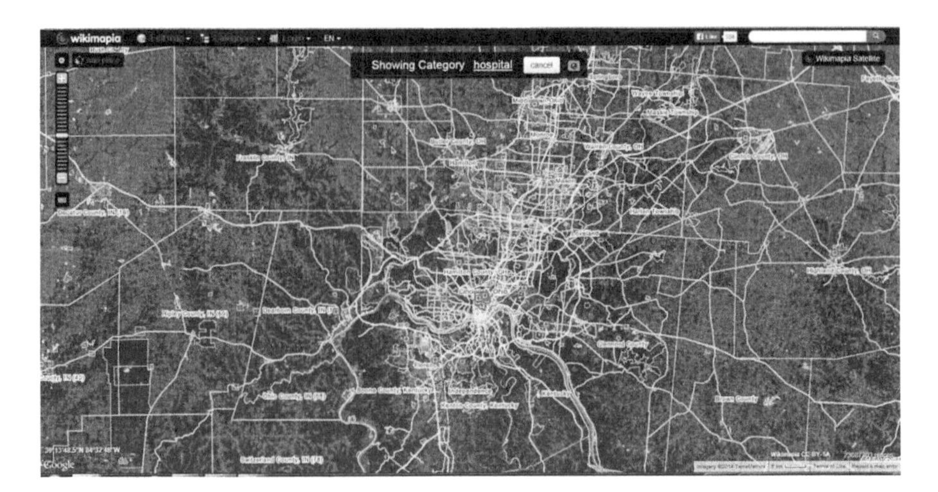

Fig. 6.6 WikiMapia is a multilingual open-content collaborative map, where anyone can create place tags and share their knowledge. Its purpose is to describe the whole world by compiling useful information about all geographical objects, organize it, and provide free access to our data for public domain (http://wikimapia.org/, Accessed 25 May 2014)

Web 2.0 map applications are able to provide basic GIS functionality to virtually anyone with a stable Internet connection. Open source solutions for Web mapping are often available at no cost, but require knowledge of digital maps, encodings and transfer protocols [29, 30]. In addition, free services – such as Google Maps, Yahoo! Maps, or Microsoft Live Maps – provide API access to programmers, and have initiated an active programming community of mashups creators. Furthermore, new formats that have been developed for the easy markup of spatial information on the Web have spurred a sharp increase in the non-professional use of online mapping tools [31, 32].

Indeed, it is this focus on geospatial functionality that is the key to success for these services. Accordingly, these recent developments have led to a split between GIS for professionals and "GIS for everyone." No longer is the production of maps and geographic information exclusive to geospatial professionals. This democratization of digital cartography is partly due to the GeoWeb, which refers to the merging of Web 2.0 with geospatial technologies and geographic information [33]. As spatial technologies and sophisticated Internet practices converge, the usage of Web 2.0 develops into a more mature type of socialization based upon shared social networks, collaborative mapping, and global information sharing. Consequently, these Internet-based geospatial services and applications have provided cartography with new features and access to the 2+ billion Internet users throughout the world [34]. This coupling of geographic information systems and hypertext systems has made the management of spatial data very efficient and user-friendly. As a result, maps have become a widely available tool for expression and participation [35, 36].

While many disease surveillance systems are implemented using the GeoWeb for the distribution and dissemination of health information, differences in operating systems, network protocols, and data models can cause problems in health information access and exchange. The integration of health data across different service systems is a major challenge in designing and implementing an effective Web-based GIS. Difference in health data inputs exist in terms of: data formats, spatial levels (e.g., point, ZIP code, county), naming conventions, terminologies, information models, and data transmission standards. In the United States, there is no central repository of health data and considerable variation exists in the formats and location requirements of the data that are reported [37]. The variability in the implementation of health standards (e.g., Health Level 7 standards) also makes it difficult to combine data from multiple health care delivery systems [38]. The sharing of health data across states or regions is uncommon, as inconsistencies across states regarding their use of geocoding references, statistical and mapping software limit the possibilities to integrate data for multi-state studies [39].

A Spatial Data Infrastructure (SDI) presents a possible solution to these problems, as it aims to promote data sharing through harmonization and standardization (thus avoiding unnecessary geospatial data duplication), at different organizational and geographical levels. SDIs provide a framework for collecting, accessing, and disseminating of geospatial data, and can enhance decision making for current problems relying on spatial data. The main components of an SDI include data providers, databases and metadata, data network, technologies, institutional arrangements, policies and standards, and end-users [40].

An illustration of the importance of a uniform SDI can be seen in the collaborations between the Association of American Geographers and the National Institutes of Health Institutes and Centers. Key leaders from these two organizations conducted a series of workshops to address the need for spatial and spatiotemporal analysis in a number of NIH biomedical and public health research centers, such as the National Cancer Institute (NCI), the National Institute of Environmental-Health Sciences (NIEHS), and the National Institute of Allergy and Infectious Diseases (NIAID), among others. During the workshop, the scientific thought leaders identified a number of data challenges, including "the lack of interoperability among proprietary systems, longitudinal variation in data collection, difficulties of sharing inadequately documented data, issues of confidentiality of location-specific data, lack of understanding of the basic concepts of spatial and spatiotemporal data and analysis, and redundancy of effort and investment" [41].

Furthermore, the report cited a number of developments in spatial and spatiotemporal data analyses and geographic computing resources that are absent in most health-science applications, such as:

"(1) the explosion of real-time, spatiotemporal data from GPS-enabled devices, distributed environmental sensor systems, satellite remote sensing, and (potentially) from geographically tagged electronic medical records; (2) development of new tools and methods for analyzing spatiotemporal data, including methods of geovisualization, dynamic spatiotemporal modeling, and modeling of human mobility at scales ranging from the everyday to the life course; and (3) advances in computing technologies, service-oriented architectures, and cyberinfrastructure that are fueling the growth of distributed and collaborative services known as the geospatial Web."

The report concluded that a comprehensive or uniform strategy to incorporate geographic context across the breadth of biomedical and public health research at NIH needs to be developed [41].

6.3 Collaborative Mapping in Public Health

The Patient Protection and Affordable Care Act (PPACA) addresses the benefits of community engagement in its provisions regarding the reductions in uncompensated care costs through an expansion of insurance coverage [42]. Effective 2 years after enactment of the PPACA, a community health needs assessment (CHNA) needs to be conducted no less than every 3 years and the reporting not-for-profit hospital must adopt a strategy to address needs identified through a CHNA and incorporate input from persons representing the broad interests of the community. The reporting hospitals must also include a description of how they are meeting identified needs through CHNA, any such needs that are not being addressed and explain why the needs are not being met in their IRS Form 990 [43].

Currently, many community benefit programs focus on counting existing activities and dollars rather than the impact of activities and the extent to which they address community needs. According to the National Institute for Healthcare Management (NIHCM), 5 % of the population accounts for almost half of total health care spending in the United States [44]. A small number of patients are consuming a large percentage of the charitable health care dollars that not-for-profit hospitals provide to their communities [45]. Having the ability to focus on the impact of resources on patients allows for improved health outcomes and potentially reduces hospital-related expenditures [46]. In addition, the targeted investments that demonstrate success in hot spot areas can also address the recent community benefits requirements enacted through healthcare reform.

Participatory hot spotting has been used to utilize chart review and data analytics for tracking patients with the highest utilization back to neighborhood and household level. A recent study identified patients at risk through hot spotting and addressed their healthcare needs in the community/neighborhood by assigning local church-based navigators to avoid fragmented care for acute episodes obtained in the emergency rooms. Study researchers applied a hot-spotting model to the Memphis, Tennessee community, and found that relatively few charitable care patients over-utilize hospital facilities (e.g., 1 of those patients had visited the emergency room 212 times in 3 years). As these patients represent the highest percentage of charitable care costs, the Memphis Model enables neighborhood community health workers to navigate those patients to more appropriate and lower level of care, and provides for regular follow-up with primary care providers in a medical home [47].

Indeed, most authorities agree that community integration is an important component of patient recovery, especially in homeless individuals who have experienced traumatic brain injuries or strokes [48]. Rehabilitation goals often include pursuing independence and participation in traditional social roles, such as education, employment, and developing social networks. Most measures of community integration rely on self-report assessments that quantify physical or social participation, but fail to capture the individual's spatial presence in the community. By documenting the daily life experiences of homeless individuals using participatory mapping and GIS, a recent study uncovered a high level of importance attributed to leisure locations (such as parks and recreational areas), as well as ongoing involvement with medical and mental health facilities [49].

To understand the benefits of parks and recreational areas in improving human health, Australian researchers utilized collaborative mapping to identify places where people engaged in various types of physical activity and where they received other benefits – such as environmental, social, and psychological health benefits [50]. Previous studies have linked urbanization and the associated loss of urban green space to poorer health and decreased quality of life for many city inhabitants [51, 52]. Upstream interventions, such as adding parks and walking trails, have been shown to have more lasting effects on physical activity levels than downstream (e.g., clinical interventions) or midstream (e.g., media campaigns) interventions [53]. The Australian researchers found that different urban park types provide different opportunities for physical activity with linear parks providing the greatest overall physical benefit [50].

Many of the examples of collaborative mapping previously in this chapter are based on the assumption that the data are publicly accessible and there are no concerns about data security. However, this concern becomes very important when mashing up sensitive healthcare information, such as medical administrative data, hospital discharge data, claims data, and medical records. As illustrated in previous chapters, the ability to integrate sensitive medical data from different sources is crucial to public health and disease surveillance activities. It is widely understood that the access to these sensitive medical databases is restricted to approved researchers and clinicians. Because it is often a requirement that the manipulation, analysis and transmission of HIPAA-compliant data be performed in a secure manner, developers and information technology vendors have begun to explore ways to provide a secure mechanism for mashing up sensitive medical information [7, 8].

References

1. O'Reilly T (2009) What is web 2.0: design patterns and business models for the next generation of software? http://oreilly.com/Web2/archive/what-is-Web-20.html. Accessed 4 May 2014
2. Hardy M (2008) Public health and web 2.0. J R Soc Promot Heal 128(4):181–189
3. Berners-Lee T, Hendler J, Lassila O (2001) The semantic web. Sci Am 284(5):34–43
4. Ankolekar A, Krotzsch M, Tran T, Vrandecic D (2007) Mashing up web 2.0 and the semantic web. In: WWW 2007. ACM, Banff
5. Boulus MNK, Scotch M, Cheung K-H, Burden D (2008) Web GIS in practice VI: a demo playlist of geo-mashups for public health neogeographers. Int J Health Geogr 7:38. doi:10.1186/1476-072X-7-38

6. Cheung K-H, Yip KY, Townsend JP, Scotch M (2008) HCLS 2.0/3.0: health care and life sciences data mashup using web 2.0/3.0. J Biomed Inform 41:694–705. doi:10.1016/j.jbi.2008.04.001
7. Kamadjeu R, Tolentino H (2006) Open source scalable vector graphics components for enabling GIS in web-based public health surveillance systems. AMIA annual proceedings/ AMIA symposium. American Medical Informatics Association, Bethesda, Maryland, p 973
8. Wang Y, Tao Z, Cross PK et al (2008) Development of a web-based integrated birth defects surveillance system in New York State. J Public Health Manag Pract 14(6):E1–E10
9. Dominkovics P, Granell C, Perez-Navarro A et al (2011) Development of spatial density maps based on geoprocessing web services: application to tuberculosis incidence in Barcelona, Spain. Int J Health Geogr 10:62. doi:10.1186/1476-072X-10-62
10. Harris TM, Weiner D, Warner TA, Levin R (1995) Pursuing social goals through participatory geographic information systems: redressing South Africa's historical political ecology. In: Pickles J (ed) Ground truth: the social implications of geographic information systems. Guilford Press, New York, pp 196–222
11. Nyerges T, Barndt M, Brooks K (1997) Public participation geographic information systems. In: Proceedings of Auto-Carto 13, Seattle, WA, American Congress on Surveying and Mapping, Bethesda, pp 224–233
12. Sieber R (2006) Public participation geographic information systems: a literature review and framework. Ann Assoc Am Geogr 96(3):491–507
13. Jankowski P, Nyerges T (2003) Toward a framework for research on geographic information-supported participatory decision-making. URISA J 15(APA I):9–17
14. Rinner C (2006) Argumentation mapping in collaborative spatial decision-making. In: Balram S, Dragićević S (eds) Collaborative geographic information systems. Idea Group Publishing, Hershey
15. Rinner C, Keßler K, Andrulis S (2008) The use of web 2.0 concepts to support deliberation in spatial decision-making. Comput Environ Urban Syst 32(5):386–395
16. MacEachren A, Brewer I (2004) Developing a conceptual framework for visually-enabled geocollaboration. Int J Geogr Inf Sci 18(1):1–34
17. MacEachren A, Cai G, Sharma R et al (2005) Enabling collaborative geoinformation access and decision-making through a natural, multimodal interface. Int J Geogr Inf Sci 19:293–317
18. MacEachren A, Guiray C, Brewer I, Chen J (2006) Supporting map-based geocollaboration through natural interfaces to large-screen displays. Cartogr Perspect 54:4–22
19. Arnott D, Pervan G (2005) A critical analysis of decision support systems research. J Inf Technol 20(2):67–87
20. Jankowski P (2009) Towards participatory geographic information systems for community-based environmental decision making. J Environ Manag 90(6):1966–1971
21. Munkvold B, Anson R (2001) Organizational adoption and diffusion of electronic meeting systems: a case study. In: Proceedings of the 2001 international ACMSIGGROUP conference of supporting group work. Boulder, ACM
22. Briggs B, Nunamaker J, Tobey D (2001) The technology transition model: a key to self-sustaining and growing communities of GSS users. In: Proceedings of the 34th Hawaii international conference on system sciences, Hawaii
23. Kolfschoten G, Duivenvoorde G, Briggs R, de Vreede G (2009) Practitioners vs facilitators a comparison of participant perceptions on success. In: Proceedings of the 42nd Hawaii international conference on system sciences. IEEE, Hawaii
24. Briggs R, Vreede G, Nunamaker J (2003) Collaboration engineering with thinklets to pursue sustained success with group support systems. J Manag Inf Syst 19(4):31–64
25. Uran O, Janssen R (2003) Why are spatial decision support systems not used? Some experiences from The Netherlands. Comput Environ Urban Syst 27:511–526
26. Zurita G, Antunes P, Baloian N et al (2008) Visually-driven decision making using handheld devices. In: Zaraté P, Belaud J, Camilleri G, Ravat F (eds) Collaborative decision-making: perspectives and challenges. Ios Press, Amsterdam, pp 257–269
27. Map Builder (2008) Map builder – rapid mashup development tool for Google and Yahoo maps. http://www.mapbuilder.net/. Accessed 25 May 2014

28. WikiMapia (2008) WikiMapia – let's describe the whole world. http://wikimapia.org/. Accessed 25 May 2014
29. MapServer (2008) MapServer – open source development environment for building spatially enabled internet applications. http://mapserver.org/. Accessed 25 May 2014
30. Deegree (2007) Deegree – free software for spatial data infrastructures. http://www.deegree.org/. Accessed 25 May 2014
31. GeoRSS (2008) GeoRSS – geographically encoded objects for RSS feeds. http://georss.org/. Accessed 25 May 2014
32. Microformats (2008) Microformats. http://www.microformats.org/. Accessed 25 May 2014
33. Herring C (1994) An architecture of cyberspace: spatialization of the internet. US Army Construction Engineering Research Laboratory, Champaign
34. Sample JT, Shaw K, Tu S, Abdelguerfi M (2008) Geospatial services and applications for the internet. Springer, New York
35. Cramptom J (2008) Cartography: maps 2.0. Prog Hum Geogr 33(1):91–100
36. Hudson-Smith A, Crooks A, Gibin M (2009) NeoGeography and web 2.0: concepts, tools and applications. J Locat Based Serv 3(2):118–145
37. National Research Council (U.S.), Committee on Research Priorities for Earth Science and Public Health (2007) Earth materials and health: research priorities for earth science and public health. National Academies Press, Washington, DC, p 188
38. Lober WB, Karras BT, Wagner MM et al (2002) Roundtable on bioterrorism detection: information system-based surveillance. J Am Med Inform Assoc 9(2):105–115
39. Gregorio DI, Samociuk H, DeChello L, Swede H (2006) Effects of study area size on geographic characterizations of health events: prostate cancer incidence in Southern New England, USA, 1994–1998. Int J Health Geogr 5:8. doi:10.1186/1476-072X-5-8
40. Coleman DJ, Nebert DD (1998) Building a North American spatial data infrastructure. Cartogr Geogr Inf Sci 25(3):151–160
41. Richardson DB, McKendry J, Goodchild M et al. (2011) Establishing an NIH-wide geospatial infrastructure for medical research: opportunities, challenges, and next steps. Association of American Geographers, Washington, DC. http://dx.doi.org/10.14433/2011.0001
42. Huntington WV, Covington LA, Center PP et al (2011) Patient Protection and Affordable Care Act of 2010: reforming the health care reform for the new decade. Pain Physician 14(1):E35–E67
43. Patient Protection and Affordable Care Act (2010) Pub L No. 111–148
44. Schoenman JA (2011) Understanding U.S. health care spending. NICHM Foundation Data Brief
45. Chazin S, Friedenzohn I, Martinez-Vidal E, Somers SA (2010) The future of US charity care programs: implications of health reform. Center for Health Care Strategies, Inc., Trenton
46. Randolph GD, Morrow JH (2013) The potential impact of the Affordable Care Act on population health in North Carolina. N C Med J 74(4):330–333
47. Cutts T, Rafalski E, Grant C, Marinescu R (2014) Utilization of hot spotting to identify community needs and coordinate care for high-cost patients in Memphis, TN. J GIS 6:23–29
48. Adair CE, Holland AC, Patterson ML et al (2012) Cognitive interviewing methods for questionnaire pre-testing in homeless persons with mental disorders. J Urban Health Bull N Y Acad Med 89(1):36–52
49. Chan DV, Helfrich CA, Hursh NC et al (2014) Measuring community integration using Geographic Information Systems (GIS) and participatory mapping for people who were once homeless. Health Place 27:92–101
50. Brown G, Schebella MF, Weber D (2014) Using participatory GIS to measure physical activity and urban park benefits. Landsc Urban Plan 121:34–44
51. Byomkesh T, Nakagoshi N, Dewan AM (2012) Urbanization and green space dynamics in Greater Dhaka, Bangladesh. Landsc Ecol Eng 8:45–58
52. Maller C, Townsend M, StLeger L, Henderson-Wilson C, Pryor A, Prosser L et al (2008) Healthy parks, healthy people: the health benefits of contact with nature in a park context. Deakin University and Parks Victoria, Burwood/Melbourne
53. Marcus B, Forsyth J (1999) How are we doing with physical activity? Am J Health Promot 14:118–124

Chapter 7
Geospatial Data Mining and Knowledge Discovery

Abstract This chapter surveys three emerging issues concerning geospatial data mining: the need to extend patient privacy protections beyond HIPAA, the use of geospatial visualization and data mining algorithms in medical geographic research, and the growth of geospatial data mining applications in public health. Geospatial data mining is the process of discovering interesting patterns in large and disparate geographic datasets so that the information is meaningful and useful to decision-makers. It involves geo-statistical algorithms, which are used for prediction, classification, and for finding interesting patterns in the data, such as associations, clusters and subgroups. A major challenge in the discipline of public health is harvesting knowledge discovery from the growing volume of data, because the discipline is a knowledge-intensive domain. Most health care applications are data-intensive and involve sophisticated data mining techniques. Since health is a geographical phenomenon, geospatial technologies play an important role in strengthening the process of epidemiological surveillance, information management and analysis.

Keywords Geospatial data mining • Geospatial visualization • Privacy preserving data mining • Modified area unit problem • Market segmentation • Target determination • Big data

7.1 Introduction

Understanding the determinants of a disease, its spread from person to person and community to community has become an increasingly complex and formidable task. The previous chapters in this volume have demonstrated that any scientific inquiry involving geographic knowledge and public health requires large volumes of multi-dimensional data. As Drs. Scholten and De Lepper have eloquently stated, "health and ill-health are affected by a variety of life-style and environmental factors, including where people live" [1]. Indeed, there are many factors – such as climate, environment, water quality and management, education, air pollution, and natural disasters – that contribute to the emergence of diseases. The characteristics of these locations (including socio-demographic and environmental exposure) offer a valuable source for epidemiological research studies on health care services and delivery.

A.J. Blatt, *Health, Science, and Place*, Geotechnologies and the Environment 12,
DOI 10.1007/978-3-319-12003-4_7

A challenge in the field of public health is knowledge discovery from the growing volume of data. Public health is a knowledge-intensive domain in which neither data gathering nor data analysis can be successful without using knowledge about both the problem domain and the data analysis process. Most health care applications are data-intensive and involve sophisticated data mining techniques. Since health is a geographical phenomenon, geospatial technologies can be useful in playing a vital role in strengthening the process of epidemiological surveillance, information management and analysis. The integration of geospatial data serves as a common platform for the convergence of multi-disease surveillance activities [2].

Data mining is the process of discovering interesting patterns in databases that are meaningful to decision-makers. It involves statistical algorithms, which result in models that can be used for prediction, classification, and for finding interesting patterns in the data, like associations, clusters and subgroups [3–5]. Geospatial data mining leverages the geographic information associated with spatial objects and finds interesting patterns, trends and relationships among them [6, 7]. Geospatial data mining discovers new and unexpected patterns, trends, and relationships embedded within large and disparate geographic data sets. Table 7.1 lists a number of geospatial data mining applications used in public health.

The analytical tools available within geographic information systems (GIS) make it possible to integrate a numerous datasets on factors that influence the spread and development of a disease. In particular, factors related to census data, economic and socio-cultural characteristics of places that affect the pattern and development of diseases. Economic factors include life expectancy, income, gender inequality, and labor mobility. Socio-cultural variables include education, religion, the ethnic composition of a population and the type and quality of the living environment. Other factors considered are the age of the epidemic, the types and availability of treatments, and sexual practices.

The goal of this chapter is to engage in discussions about three emerging issues regarding geospatial data mining. Privacy protections – first discussed in Chap. 5 of this volume – is worth revisiting, as it can be extend beyond the *Safe Harbor* standard described in the Privacy Rule in the Health Insurance Portability and Accountability

Table 7.1 Geospatial data mining applications in public health

(1) Uncovering the geographical distribution and variation of diseases.
(2) Analyzing the spatial trends in disease and interventions over time.
(3) Forecasting epidemics.
(4) Identifying gaps in immunizations.
(5) Mapping populations at risk and stratifying risk factors.
(6) Documenting health care needs of a community and assessing resource allocations.
(7) Planning and targeting interventions.

Act of 1996 (HIPAA), especially in data mining and knowledge discovery applications. Next, the combination of geospatial visualization and data mining algorithms provides a power tool "to help a user to get a feeling for the data, to detect interesting knowledge, and to gain a deep visual understanding of the data set" [8]. Finally, this chapter will end with a discussion of the numerous geospatial data mining applications in health care – from market segmentation and target development in the pharmaceutical industry to environmental health risk assessments and health disparities and outcomes research.

7.2 Patient Protection Beyond De-identification

Privacy is, and will continue to be, a great concern to patients and medical providers. Individuals can be deeply affected by the inappropriate disclosure of their medical information. The range of negative outcomes can involve the denial of health insurance, or the loss of a new job bid resulting from leaked information word about an illness or condition. As stated in Chap. 5, HIPAA regulations in the United States protect health information from being shared or published. Protected health information (PHI), according to HIPAA, is: "Any information, whether oral or recorded in any form or medium" that is "created or received by a health care provider, health plan, public health authority, employer, life insurer, school or university, or health care clearinghouse" and "relates to the past, present, or future physical or mental health or condition of an individual; the provision of health care to an individual; or the past, present, or future payment for the provision of health care to an individual" [9]. This wording effectively places any identifiable part of a medical record under the protection of HIPAA.

HIPAA provides three standards for health information disclosure without seeking patient authorization: the *Safe Harbor*, the *Limited Dataset,* and the *Statistical Dataset* standards. *Safe Harbor* specifies very precise rules to remove PHI when medical data is released for secondary purposes. It requires the deletion of 18 variables (as described in Chap. 5, Table 5.1). The *Limited Dataset* standard allows the dataset to contain dates (like date of birth) and geographic information (except street address). This standard is not for publicly released data, and requires the user to sign a data use agreement. The data use agreement typically sets strict requirements about data storage and the personnel who would have access to the data. It also requires the user to destroy the data as soon as it fulfilled its purpose. The *Expert Determination* standard requires that an expert (with appropriate knowledge and experience with generally accepted statistical and scientific methods in the practice of data de-identification) certifies that there is a very small risk that the information could be used by the recipient to identify the individuals in the dataset, alone or in combination with other reasonably available information. The person certifying statistical de-identification must document the methods used as well as the result of the analysis that justifies the determination.

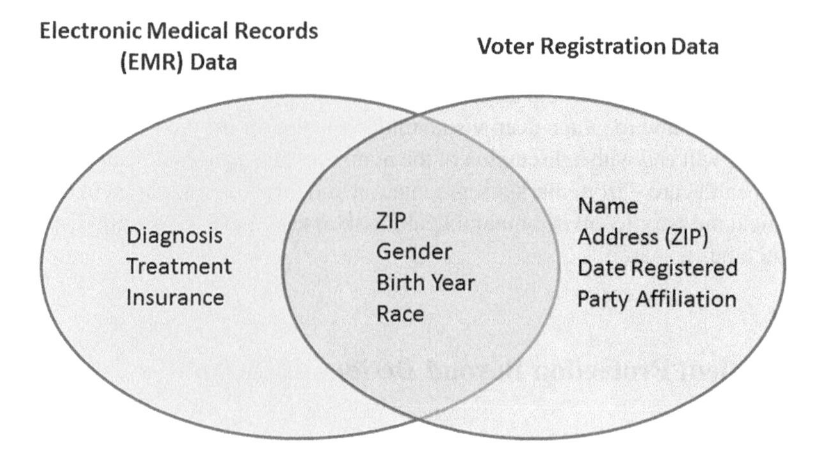

Fig. 7.1 Linking sensitive and private dataset with a public dataset

Most medical providers in the United States follow the *Safe Harbor* standard in the when releasing data. Unfortunately, removing identifiers does not completely ensure protection against intrusions of patient privacy. As others have demonstrated, it is possible to link the de-identified privacy sensitive dataset with publicly available data and discover the identity of certain individuals [10, 11]. Figure 7.1 shows the linkage of a de-identified medical dataset with voter registration data. Attributes such as ZIP code, sex, birth year and race are available in both datasets. Table 7.2 shows sample records in each datasets. Although the obvious identifiers are removed from the medical dataset, it is possible that a combination of non-identifier attributes is unique and can be used to identify an individual. These attributes – ones that can uniquely identify individuals and might be available in public datasets – are called quasi-identifiers.

In the hypothetical example, there was only one person in both datasets who was a female, had a zip code of 405XX, was born in 1962 and whose race was Native Hawaiian and Other Pacific Islander. By linking the two datasets on these variables, it is possible to discover the person's identity and medical conditions. This example demonstrates that simply removing identifiers from a dataset does not solve the de-identification problem, as identity disclosure is still possible.

As demonstrated above, the possibility of identifying individuals in publicly released datasets remains a real privacy concern. There is a rich body of work (in the field of privacy preserving data mining) on the implementation and application of methods to prevent these types of disclosures [12–15]. However, most of this research involves datasets with a single sensitive attribute, and it not suitable for medical datasets that often contain multiple sensitive attributes (e.g., diagnosis and treatment). Another problematic issue is that medical data providers lack knowledge and awareness of how the data will be used by the researchers who will ultimately analyze the data. Data providers need to prepare anonymized datasets using customized

Table 7.2 Sample records in hypothetical example

(A) Medical data

ZIP(3)	Sex	BirthYear	Race	Diagnosis	Treatment	Insurance
405XX	F	1924	African American	Cancer	Chemo	HMO
405XX	M	1972	Native Hawaiian and Other Pacific Islander	Hepatitis C	Antiviral	Medicaid
405XX	F	1946	White	HIV	ETR	Not Insured
405XX	M	1955	African American	Multiple Sclerosis	Mitoxantrone	Medicare

(B) Voter registration data

Name	ZIP	Sex	BirthYear	Race	DateRegist	Party
Jane Doe	40504	F	1945	White	12/21/2001	GOP
Joe Fogle	40506	M	1972	Native Hawaiian and Other Pacific Islander	6/9/1987	GOP
Mark Right	40521	M	1936	White	3/14/1990	DEM
Maria Left	40504	F	1972	African American	11/8/2007	GOP
Anna Parks	40509	F	1967	African American	8/30/2010	DEM

(C) Linked data

Name	ZIP	Sex	BirthYear	Race	Diagnosis	Treatment	Insurance
Joe Fogle	40506	M	1972	Native Hawaiian and Other Pacific Islander	Hepatitis C	Antiviral	Medicaid

methods that preserves the greatest utility that is specific to the needs of the recipient of the data – as prescribed in existing research on utility-based privacy preserving data mining [16–18].

The issues raised above result from the rapid growth of vast computerized datasets at unprecedented rates. These datasets come from various sectors, e.g., business, education, government, scientific community, Internet, or one of many readily available off-line and online data sources in the form of text, graphics, images, video, audio, animation, hyperlinks, and markups. In addition, they are continuously increasing and are amassed in both the attribute depth and the scope of instance objects every time. As much of this data is geo-referenced, sources of data important to public health investigations may come from natural resource investigations, surveying and mapping, astronomical data, satellites, and spacecraft images. They include not only positional data and attribute data, but also temporal relationships among spatial entities (Fig. 7.2) [19]. Spatial data structure is inherently more complex than the tables in an ordinary relational database. In addition to tabular data, there are raster images and vector graphics, whose attributes are not explicitly stored in the database. Fortunately, contemporary GIS functionalities and powerful computer processors are designed to make full use of these vast and multi-dimensional spatial datasets.

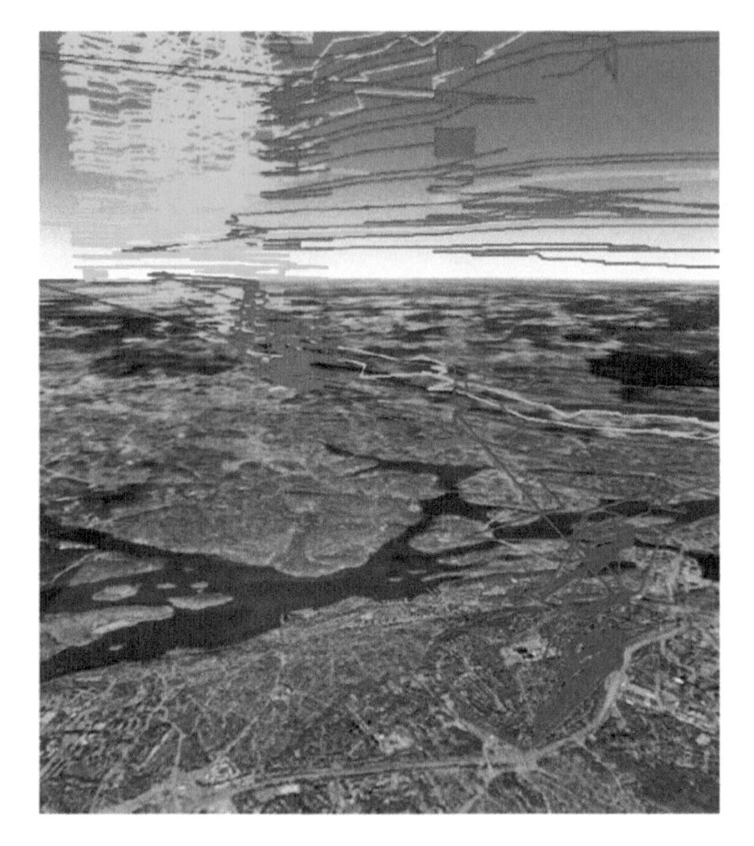

Fig. 7.2 Time-space representation of four paths, each represented by a different color. The time dimension is represented by altitude (From: Sanches et al. [19])

7.3 Exploring Geospatial Data Through Visualization

Data visualization has been used as an important tool to gain insights into health care data sets, which are typically multivariate, discrete, and at different temporal and spatial levels. Displaying geographically distributed health related data in a chloropleth map using sequential color scheme has been proven useful for data driven knowledge discovery, for the non-technical user [20–22]. Although many previous works have employed different desktop applications designed for multivariate analyses to study health care datasets [23, 24], a web-based tool for health care data visualization has been found to provide better public data, access, and usability [25–28].

The ability to visualize multiple variables on the map and compare them using tables and charts at the same time can provide valuable insights. While GIS has been found to be a useful method to visualize the health care data, there also have been concerns that using interactive maps without any other form of data

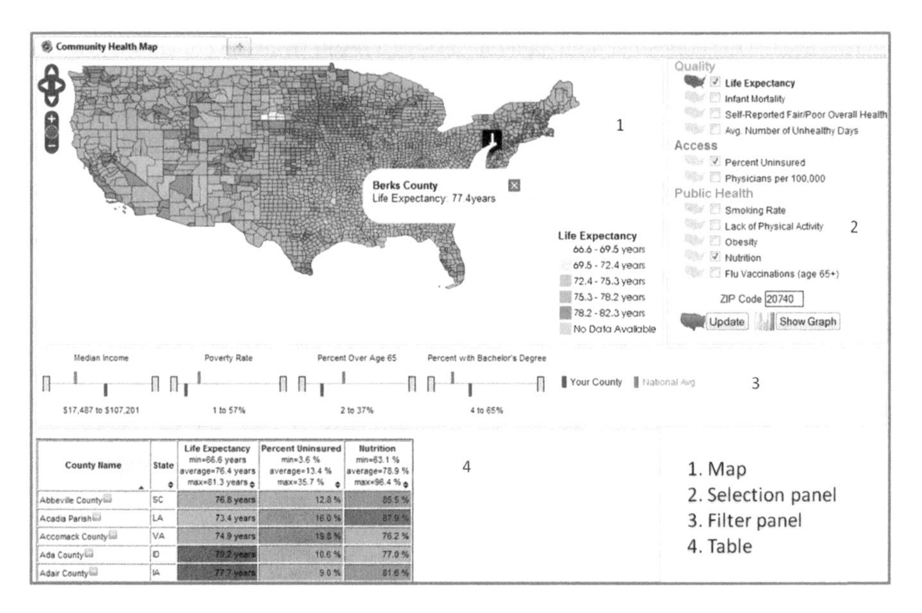

Fig. 7.3 The Community Health Map interface consists of a map (*top*), hinted double-sided sliders for map and table filtering (*middle*), a color coded table (*bottom*), and the selection panel (*right*). The selection panel allows users to click the link to render a layer on the map, or check boxes to display a table with the selected variables (From: Sopan et al. [33])

visualization might be subject to misinterpretation [29–31]. Although a conceptual frame-work and potential capabilities of such a tool have been discussed, there are very few applications in existence [32, 33]. An example of a user-friendly data visualization tool, Community Health Map provides web-based data discovery functionality that is designed to visualize health care data at the county or Hospital Referral Region (HRR) level (Fig. 7.3). The tool is designed for users with experience in the health care industry and it does not require programming experience. Users are able to visualize health performance, access, and quality indicators for their county or HRR; filter the geographic regions by demographic characteristics (such as age, income, poverty, or education), for a comparison to the national average; and produce visual presentations that allow policy makers to take action to improve the health and quality of health care in their regions and streamline healthcare spending [33].

In cases where understanding the spatial epidemiology of a disease depends upon the level of aggregation of the spatial data [34, 35], a number of exploratory data visualization methods have been developed to handle complex spatial data and the Modified Area Unit Problem. Disease locations are often reported in either point or aggregated form. In the former case, the location is given by a pair of coordinates (x; y), e.g. referring to Cartesian coordinates or to longitude and latitude. Alternatively, locations can be represented in aggregated form as a subset of the study region, called the aggregation unit, typically an administrative level such as ZIP code or census tract. This coarser spatial data on cases can result from the collection process or might have been intentionally aggregated due to privacy. Measuring the effect of spatial aggregation is also termed the effect of scale or effect of discretization, and is a particular aspect of the well-known and well-documented Modifiable Areal Unit Problem [36].

Numerous researchers have studied the relationship between an individual's location and the acquisition of disease, often developing novel exploratory disease mapping methods to estimate a change in the spatial distribution of cases from a reference distribution (which represents a residual spatial risk surface after all known risk factors have been accounted for) [37]. In addition, several studies have utilized non-parametric disease distance-based mapping methods to study the effects of spatial aggregation across different geographic scales using simulated spatial data under controlled conditions. Although spatial aggregation produces an expected degradation of disease mapping performance, the characteristics of this degradation has been found to vary depending on the interaction between the geographic extent of the higher risk area and the level of aggregation – i.e., higher risk areas dispersed across several units did not suffer as greatly from aggregation [34, 35, 38, 39]. Because concerns about privacy related to electronically collected spatial data are an important part of the public health discourse, a discussion of the scale effect and spatial risks warrants a careful and systematic methodological study of the compromises imposed by coarsened, or masked, data repositories.

7.4 Geospatial Data Mining Applications in the Healthcare Industry

Historically speaking, the healthcare industry has generated vast data warehouses of information, driven by record keeping, compliance and regulatory requirements, and patient care [40]. While most of this information is stored in hard-copy form, the current trend is towards the rapid digitization of these large troves of data. Driven by mandatory requirements and the potential to improve the quality of healthcare delivery meanwhile reducing the costs, these massive quantities of data (known as 'big data') hold the promise of supporting a wide range of medical and healthcare functions, including clinical decision support, disease surveillance, and population health management [41]. Data from the U.S. healthcare system reached an estimated 150 exabytes, in 2011 alone. Kaiser Permanente, the California-based health network, which has more than 9 million members, is believed to have between 26.5 and 44 petabytes of potentially rich data from EHRs, including images and annotations [42].

By definition, big data in healthcare refers to electronic health data sets so large and complex that they are difficult to manage with traditional software and hardware equipment. Big data is overwhelming because of its volume, the diversity of data types – including geospatial data – and the speed at which it must be managed. Big data includes clinical data from computerized physician order entry and clinical decision support systems (e.g., physician's written notes and prescriptions, medical imaging, laboratory, pharmacy, insurance, and other administrative data); patient data in electronic health records (EHRs); machine generated/sensor data, such as from monitoring vital signs; social media posts, including Twitter feeds, status updates on Facebook and other platforms, and web pages; and less patient-specific information, including emergency care data, news feeds, and articles in medical journals [42, 43]. It should be noted that most of these data sources, by nature, have a geographic component.

By discovering associations and understanding the geographic patterns and trends within the data, geospatial data mining has the potential to improve care, save lives and lower costs. When geospatial healthcare data is synthesized and analyzed, healthcare providers and other stakeholders in the healthcare delivery system can develop more thorough and insightful diagnoses and treatments, resulting, one would expect, in higher quality care at lower costs and in better outcomes. The potential for geospatial data mining in healthcare to lead to better outcomes exists in many facets of health care services and delivery. For instance, by analyzing a patient's disease ecology and the cost and outcomes of care, one could custom tailor the most clinically and cost effective treatments and offer analysis and tools, thereby influencing provider behavior and improving patient outcomes. In addition, applying advanced geo-statistical analyses to patient profiles (e.g., segmentation and environmental hazards modeling) could proactively identify individuals who would benefit from preventative care or lifestyle changes. Furthermore, broad-scale spatial epidemiological studies can identify predictive environmental events and support population-specific prevention initiatives. Finally, the development and deployment of geographically enabled mobile apps can help patients manage their care, assess health and environmental risks, locate providers and improve their health [44–47].

In conclusion, leveraging integrated geospatial information has the potential to transform the way healthcare providers use sophisticated technologies to gain insight from their clinical and other data repositories and to make informed decisions. Several challenges associated with the rapid and widespread use of geospatial data mining across the healthcare organization and the healthcare industry must be addressed. As geospatial technologies become more mainstream, issues such as guaranteeing privacy, safeguarding security, establishing standards and governance, and continually improving the tools and technologies will garner attention. Geospatial data mining applications in the healthcare industry are at a nascent stage of development, but the current and rapid advances in cloud-based server platforms and web technologies can accelerate their maturing process.

References

1. Scholten HJ, De Lepper MJC (1991) The benefits of the application of Geographic information systems in public and environmental health. WHO Stat Q 44(3):160–171
2. WHO (1999) Geographical information systems (GIS): mapping for epidemiological surveillance. Wkly Epidemiol Rec. Nilmini Wickramasinghe, and Eliezer Geisler, 74(34):281–285

3. Kuo RJ, Li SY, Shi CW (2007) Mining association rules through integration of clustering analysis and ant colony system for health insurance database in Taiwan. Expert Syst Appl 33:794–808
4. Lavrac N, Bohanec M, Pur A et al (2007) Data mining and visualization for decision support and modeling of public health-care resources. J Biomed Inform 40(4):438–447
5. Mullinsa IM, Siadatya MS, Lymana J (2005) Data mining and clinical data repositories: insights from a 667,000 patient data set. Comput Biol Med 36:1351–1377
6. Shekhar S, Vatsavai R (2003) Techniques for mining geospatial databases, as chapter 22. In: Ye N (ed) Handbook of data mining. LEA Publishers, Hillsdale
7. DemSar U (2007) Investigating visual exploration of geospatial data: an exploratory usability experiment for visual data mining. Comput Environ Urban Syst 31(5):551–571
8. Beilken C, Spenke M (1999) Visual interactive data mining with InfoZoom – the Medical Data Set. In: Proceedings of the 3rd European conference on principles and practice of knowledge discovery in databases, PKDD 1999, Prague, Czech Republic
9. Health Insurance Portability and Accountability Act of 1996. Pub. L. 104–191 (Aug. 21, 1996)
10. Sweeney L (2002) K-anonymity: a model for protecting privacy. Int J Uncertainty Fuzziness Knowl Based Syst 10(5):557–570
11. Samarati, P, Sweeney L (1998) Protecting privacy when disclosing information: k-anonymity and its enforcement through generalization and suppression. Technical report. SRI International
12. LeFevre K, DeWitt DJ, Ramakrishnan R (2005) Incognito: efficient full-domain k-anonymity. In: Proceedings of the 2005 ACM SIGMOD international conference on Management of data, Maryland, New York, pp 49–60. doi:10.1145/1066157.1066164
13. Menon S, Sarkar S (2006) Exploiting problem structure to efficiently sanitize very large transactional databases. In: The 16th workshop on information technology and systems, Milwaukee, Wisconsin
14. Samarati P (2001) Protecting respondents' identities in microdata release. IEEE Trans Knowl Data Eng 13(6):1010–1027
15. Xiao X, Tao Y (2006) Anatomy: simple and effective privacy preservation. In: Proceedings of the 32nd international conference on very large data bases, Seoul, Korea, pp 139–150
16. Xu J, Wang W, Pei J et al (2006) Utility-based anonymization for privacy preservation with less information loss. SIGKDD Explor Newsl 8(2):21–30. doi:10.1145/1233321.1233324
17. Xu J, Wang W, Pei J et al (2006) Utility-based anonymization using local recoding. In: Proceedings of the 12th ACM SIGKDD international conference on knowledge discovery and data mining, ACM Press, New York, pp 785–790. doi:10.1145/1150402.1150504
18. Zhang Q, Koudas N, Srivastava D et al (2007) Aggregate query answering on anonymized tables. In: 2007 IEEE 23rd international conference on data engineering, Istanbul, Turkey, pp 116–125. doi:10.1109/ICDE.2007.367857
19. Sanches P, Svee E-O, Bylund M et al (2013) Knowing your population: privacy-sensitive mining of massive data. Netw Commun Technol 2(1):34–51. doi:10.5539/nct.v2n1p34
20. MacEachren AM, Brewer CA, Pickle LW (1998) Visualizing georeferenced data: representing reliability of health statistics. Environ Plan A 30(9):1547–1561
21. Koua EL, Kraak MJ (2004) Geovisualization to support the exploration of large health and demographic survey data. Int J Health Geogr 3(1):12
22. Tominski C, Schulze-Wollgast P, Schumann H (2008) Visual methods for analyzing human health data. In Wickramasinghe N, & Geisler E (eds): Encyclopedia of healthcare information systems. Medical Information Science Reference, Hershey, Pennsylvania, pp 1357–1364
23. Keahey TA (1998) Visualization of high-dimensional clusters using nonlinear magnification. Vis Data Expl Anal VI 3643:228–235
24. Madigan EA, Curet OL (2006) A data mining approach in home healthcare: outcomes and service use. BMC Health Serv Res 6(1):18. doi:10.1186/1472-6963-6-18
25. Verdegem P, Verleye G (2009) User-centered e-government in practice: a comprehensive model for measuring user satisfaction. Gov Inf Q 26(3):487–497
26. Gil-Garcia JR, Pardo TA (2005) E-Government success factors: mapping practical tools to theoretical foundations. Gov Inf Q 22(2):187–216

27. Fedorowicz J, Dias MA (2010) A decade of design in digital government research. Gov Inf Q 27(1):1–8
28. Donker-Kuijer MW, de Jong M, Lentz L (2010) Usable guidelines for usable websites? An analysis of five e-government heuristics. Gov Inf Q 27(3):254–263
29. Goodman DC, Wennberg JE (1999) Maps and health: the challenges of interpretation. J Public Health Manag Pract 5(4):xiii–xvii
30. Villalon M (1999) GIS and the internet: tools that add value to your health plan. Health Manag Technol 20(9):16–18
31. Castronovo D, Chui KKH, Naumova EN (2009) Dynamic maps: a visual-analytic methodology for exploring spatio-temporal disease patterns. Environ Heal 8:61. doi:10.1186/1476-069X-8-61
32. Lu X (2005) A framework of web GIS based unified public health information visualization platform. Comput Sci Appl ICCSA 3482:265–268
33. Sopan A, Noh AS-I, Karol S et al (2012) Community Health Map: a geospatial and multivariate data visualization tool for public health datasets. Gov Inf Q 29:223–234
34. Jeffery C, Ozonoff A, White LF et al (2009) Power to detect spatial disturbances under different levels of geographic aggregation. J Am Med Inform Assoc 16:847–854
35. Jones SG, Kulldorff M (2012) Influence of spatial resolution on space-time disease cluster detection. PLoS ONE 7(10):e48036. doi:10.1371/journal.pone.0048036
36. Gotway CA, Young LJ (2002) Combining incompatible spatial data. J Am Stat Assoc 97:632–649
37. Wakefield J, Kelsall J, Morris S (2000) Clustering, cluster detection, and spatial variation in risk. In: Elliott P, Wakefied JC, Best NG, Briggs DJ (eds) Spatial epidemiology: methods and applications. Oxford University Press, Oxford, pp 128–152
38. Jeffery C, Ozonoff A, Pagano M (2014) The effect of spatial aggregation on performance when mapping a risk of disease. Int J Health Geogr 13:9. doi:10.1186/1476-072X-13-9
39. Ozonoff A, Jeffery C, Pagano M (2009) Multivariate disease mapping. In: Proceedings of the American Statistical Association, Biometrics Section [CD-ROM] ASA
40. Raghupathi W (2010) Data mining in health care. In: Kudyba S (ed) Healthcare informatics: improving efficiency and productivity. CRC Press/Taylor & Francis Group, Boca Raton, pp 211–223
41. Fernandes L, O'Connor M, Weaver V (2012) Big data, bigger outcomes. J AHIMA 83(10):38–42
42. IHTT (2013) Transforming health care through big data: strategies for leveraging big data in the health care industry. http://ihealthtran.com/iHT2_BigData_2013.pdf. Accessed 5 Apr 2014
43. Bian J, Topaloglu U, Yu F et al (2012) Towards large-scale Twitter mining for drug-Related adverse events. In: Proceedings of the 2012 international workshop on smart health and wellbeing, pp 25–32. doi:10.1145/2389707.2389713
44. Savage N (2012) Digging for drug facts. Commun ACM 55(10):11–13
45. LaValle S, Lesser E, Shockley R et al (2011) Big data, analytics and the path from insights to value. MIT Sloan Manag Rev 52:20–32
46. Courtney M (2013) Puzzling out big data. Eng Technol 7(12):56–60
47. Raghupathi W, Raghupathi V (2014) Big data analytics in healthcare: promise and potential. Health Inf Sci Syst 2:3. doi:10.1186/2047-2501-2-3

Part IV
Geography in Medicine

Chapter 8
Geographic Medicine

Abstract This chapter highlights a sub-discipline of medicine, known as geographic medicine, to describe how human movements contribute to the transmission of parasites on spatial scales that exceed the limits of its natural habitat. Traditionally, public health programs have focused on the health of populations, whereas the practice of medicine has focused on the health of individuals. It should be noted, however, that the population health management owes much to the effective delivery of clinical care. This chapter demonstrates how public health is intimately linked to patient care through human movement. Nearly a century ago, people typically did not develop a disease where it is contracted or even close to that place. Today, daily travel is a common way of life in modern metropolitan areas. Large, localized mosquito populations in areas that people visit regularly may be both reservoirs and hubs of infection, even if people only pass through those locations briefly. By examining of the role of human movement across different scales, this chapter examines how public health communities can use information on pathogen transmission to increase the effectiveness of disease prevention programs and clinical care.

Keywords Geographic medicine • Human movement • Vector-borne diseases • Pathogen transmission • Globalization • Mass air transportation • Public health • Clinical care

8.1 Introduction

In 2008, the United States' Institute of Medicine convened an Expert Committee on the U.S. Commitment to Global Health and reaffirmed the notion that local health and local health care are linked to sources of disease and disability occurring elsewhere in the world [1]:

> Global health is the goal of improving health for all people in all nations by promoting wellness and eliminating avoidable diseases, disabilities, and deaths. It can be attained by combining clinical care at the level of the individual person with population-based measures to promote health and prevent disease. This ambitious endeavor calls for an understanding of health determinants, practices, and solutions, as well as basic and applied research concerning risk factors, disease, and disability.

Traditionally, public health has focused on the health of populations, whereas the practice of medicine has focused on the health of individuals. However, it should be

A.J. Blatt, *Health, Science, and Place*, Geotechnologies and the Environment 12,
DOI 10.1007/978-3-319-12003-4_8

noted that the population health management owes much to the effective delivery of clinical care and how medical activities affect it. In 2013, an entire issue of the *New England Journal of Medicine* was devoted to the problems and solutions of global health. In its Introduction, the editors wrote [2]:

> Diseases of global importance, such as injuries, noncommunicable diseases, and mental health, can be partitioned according to their differential geographic and temporal effects… Global forces such as globalization and climate change, as well as personal behavior involving risk factors such as tobacco use, excessive alcohol consumption, and poor diet, affect health in all countries.

The last three chapters of this book will illustrate the importance of geography in the practice of medicine, both on a local and global scale. The current chapter will highlight the role of geographic forces in the field of travel medicine (also known as geographic medicine) and emphasize the role of geographic knowledge and thought in a study of disease that views humans has vectors and hosts. Chapter 9 will introduce a relatively new discipline, called geospatial medicine, that utilizes the advances in geospatial data and technology to uncover the genetic, social, and environmental effects of disease as it occurs throughout human development. This is an exciting new field in which there is a tremendous potential for geographers to take center stage in answering some of the hard research questions that scientists have been grappling with for a long time. Lastly, Chap. 10 puts the finishing touches on the new portrait of medical geography that we have been painting, and frames it in a more centralized role on the national stage of the U.S. health care reform. A main objective of this book is to present a new model of patient care that emphasizes the patient's geographic and medical history, against the backdrop of contemporary globalization, while taking advantage of the current advances in geospatial data and technologies.

The current chapter will describe the main principles of travel medicine, and emphasize how human movements contribute to the transmission of parasites on spatial scales that exceed the limits of its natural habitat. Studies have shown that the ability to identify the sources (origins) and sinks (destinations) of imported infections due to human travel and locating the high-risk sites of parasite importation could greatly improve the control and prevention programs [3, 4]. For instance, a group of researchers have combined mobile phone data with a high-resolution malaria prevalence map to analyze the regional travel patterns of nearly 15 million individuals over the course of a year in Kenya, in the context of parasitic dispersals. In this way, they were able to identify and map the sources and sinks of human and parasite travel in this country (Fig. 8.1) [5].

To describe the challenges of travel medicine on human terms – in 2013, an estimated 1.1 billion travelers crossed international borders, including an estimated 28.5 million U.S. travelers [6, 7]. An increase in the number of international travelers and local commuters has caused a surge of global and regional epidemics in the past several decades [8]. Travel-related risks include infections from food, vectors, and bodily fluids. Most travelers (20–70 %) report health problems while traveling, but many do not seek pre-travel advice – such as vaccinations, prophylactics, and therapeutic medications [9, 10]. Moreover, the risk of travel-related diseases is 2.3 times higher in those with an underlying medical conditions than in health individuals [10]. Given the unique health situations of travelers, there is a need for better characterization

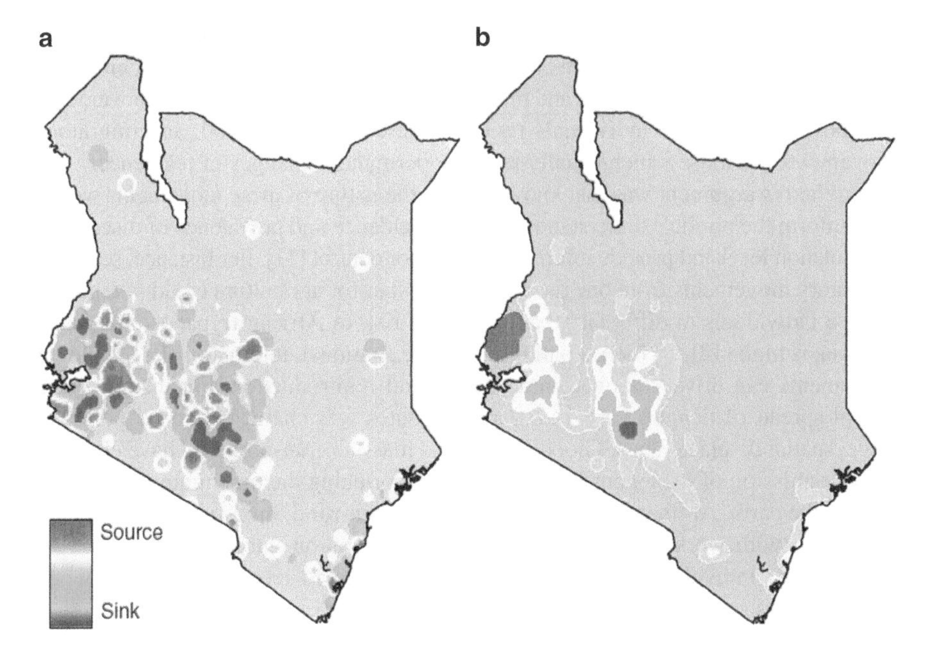

Fig. 8.1 Sources and sinks of people (**a**) and parasites (**b**). Kernel density maps showing ranked sources (*red*) and sinks (*blue*) of human travel and total parasite movement in Kenya, where each settlement was designated as a relative source or sink based on yearly estimates (From: Figure 3 in Wesolowski et al. [5])

of the geographical and environmental risk factors underlying travel-related illnesses. To address these concerns, this chapter describes how human travelers who are infected can serve as an important route for the transmission of the virus from their place of origin to their place of destination. By understanding the effects of human movement on disease transmission, researchers can appropriately identify high-risk areas for effective intervention and control.

8.2 Importance of Human Movements in Disease Transmission

Historically, epidemiologists have viewed human movement from two main perspective. The first perspective is from that of the populations of susceptible hosts moving into high-risk areas. The second perspective is from that of populations of infected hosts moving into susceptible populations. Movements of infected hosts across different spatial scales affect pathogen transmission in a variety of ways. Noted historian and geographer R. Mansell Prothero published one of the first studies describing the role of human movements in epidemiology based on his experience in Africa, in 1977. Drawing on the geographic literature concerning diffusion and migration processes, he discussed the relevance of these movement patterns to public health in his seminal paper published in the *International Journal of Epidemiology*.

He carefully outlined the differences between circulatory and migratory movements and categorized these movements by their spatial scale (i.e., a rural-urban gradient) and temporal scale (i.e., the time and timing of displacements). Circulatory movements are those in which the individuals return home after some period, and migratory movements are those which usually result in permanent changes of residence.

Prothero's argument was that knowledge of the nature of these movements would help inform the public's understanding of the incidence and prevalence of disease on a population level and provide informed options for control [11]. For instance, seasonal migratory movements from one rural area to another for agriculture could potentially expose individuals to different areas where the risk of African trypanosomiasis, or malaria, is high [12]. At broad spatial scales (e.g., national, international), individual movements can drive pathogen introduction and reintroduction. For example, the global spread of dengue virus via shipping routes was characterized by periodic, large, spatial displacements. Globalization and mass air transportation have changed the transmission of pathogens by dramatically shortening the time required to travel around the earth. At finer scales (e.g., regional, urban-rural, intra-urban), movement associated with work, recreation, and transient migration into high-risk areas not only lead to individual infection, but also contribute to local transmission when infected hosts return home and infect other individuals (Fig. 8.2) [13].

Indeed, vector-borne diseases place an enormous burden on public health and require efficient control strategies that are developed through an understanding of the origin (or sources) of infections and the relative importance of human movement at different scales. A number of social and environmental factors – such as human population density, regional settlement patterns, population movements, precipitation, and other weather-related factors – contribute to local and regional transmission dynamics [14]. Human movement – which determines exposure to vectors – is a key aspect of vector ecology that is poorly understood. This is due to the variations in exposure based on individual host movement that can strongly influence the pathogen's transmission dynamics [15].

The types of movement most relevant for exposure will depend on site-specific differences, the ecology of the arthropod vector, human behavior, and the relative scale of host and vector movement. For instance – although fine-scale host movements are not important to pathogens transmitted by vectors that are able to move long distances in search of a host – these fine-scale host movements are very important for pathogens transmitted by sessile vectors. *Aedes aegypti* is the principal vector of dengue virus. It bites during the day, disperses only short distances, and is heterogeneously distributed within urban areas. Humans, on the other hand, move frequently and allocate different amounts of time to multiple locations on a regular basis. Not only does this influence the individual risk of infection with dengue virus but it also influences overall patterns of transmission [16]. In addition, commuting and non-commuting patients have different diffusion patterns and determinants in a dengue epidemic. Non-commuters (e.g., elderly adults and housewives) may initiate a local epidemic, whereas commuters carrying the virus to geographically distant areas can cause a large-scale epidemic [17].

Dengue is a global threat and is endemic or epidemic in almost every country located in the tropics (Fig. 8.3). While new tools (such as vaccines, antiviral drugs

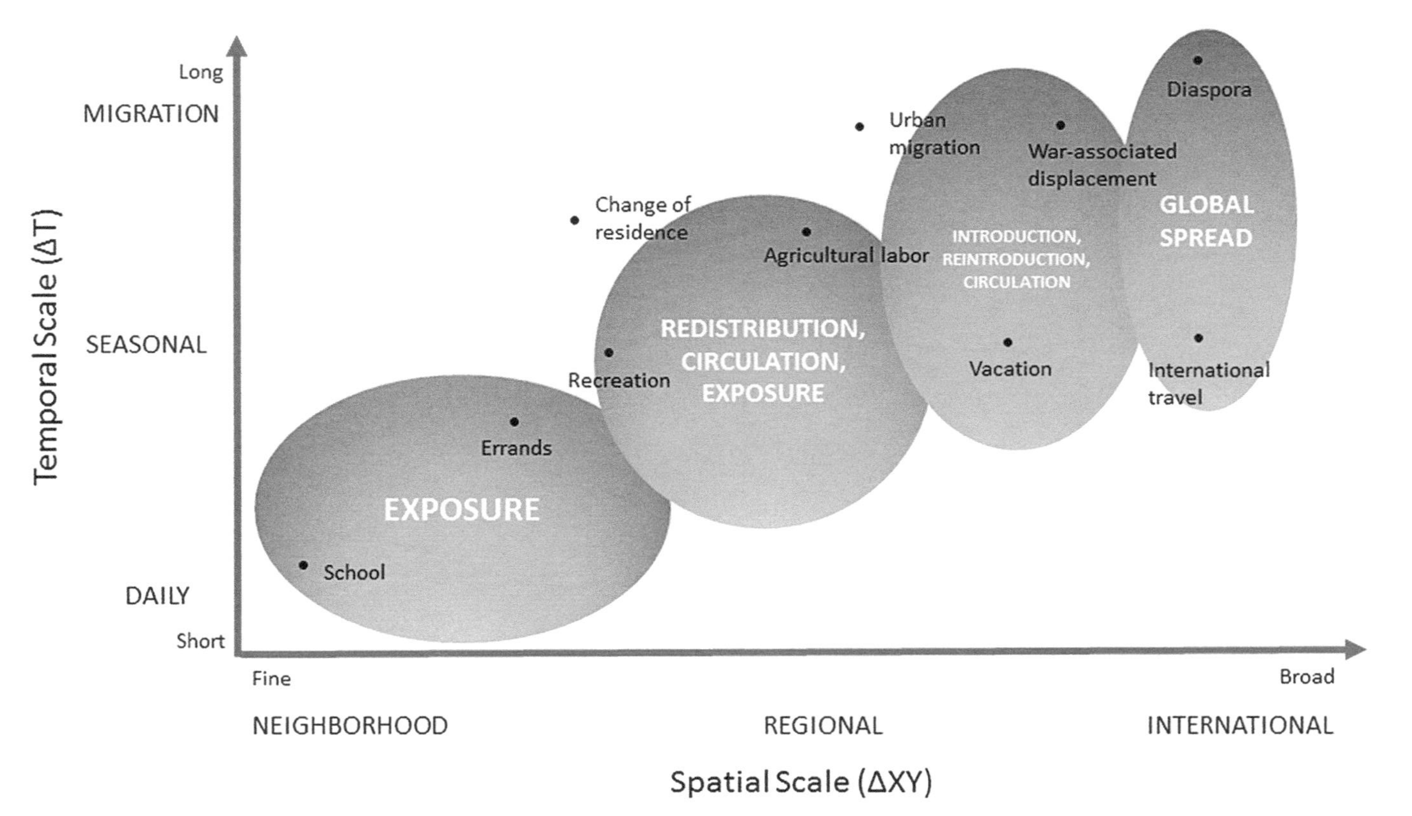

Fig. 8.2 A framework for human movements and their relevance to vector-borne pathogen transmission. Movements are characterized in terms of their spatial and temporal scale, which are defined in terms of physical displacement (ΔXY) and time spent (ΔT, frequency and duration). Generally, movements of greater spatial displacement involve more time (Adapted from: Figure 1 in Stoddard et al. [15])

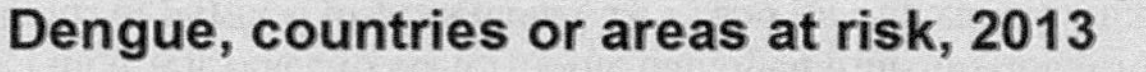

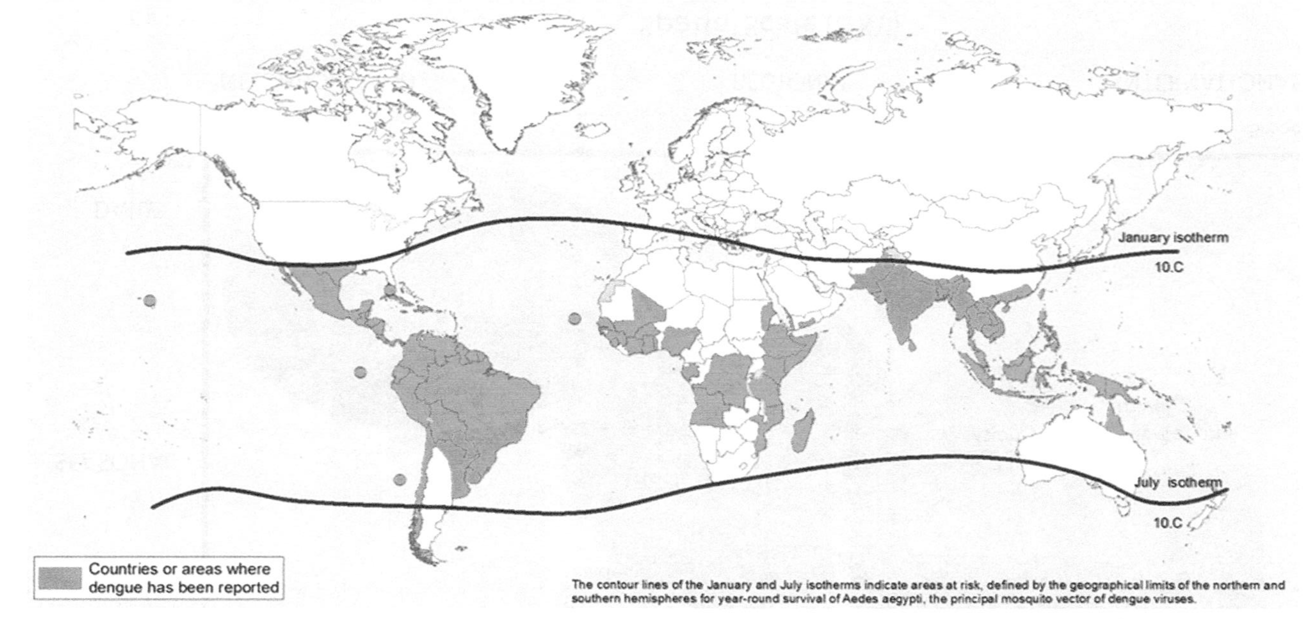

Fig. 8.3 Counties and areas at risk of dengue transmission, 2008 (Source: World Health Organization Map Production: Health Systems and Information Systems (HSI). http://gamapserver.who.int/mapLibrary/Files/Maps/Global_DengueTransmission_ITHRiskMap.png)

and improved diagnostics) are being developed, better use should be made of the interventions that are currently available. Along with comprehensive tracking of commuting cases, the concomitant rapid notification and diagnosis of non-commuting cases can enable the appropriate interventions and faster response times, thus preventing subsequent large-scale epidemics.

8.3 Geographic Considerations in Human Movement

As illustrated in the previous sections of this chapter, human movement is a largely ignored variable in the study and diagnosis of pathogen exposures. There are few, if any, well-established clinical guidelines that utilize a person's place history in the diagnosis of infected individuals, despite the fact that disease management protocols can be made more effective and efficient by targeting the sources or agents of transmission. The study of human movement is critical to identifying high-risk factors (e.g., host-preferences) because these factors are always conditioned by exposure rates which are, in turn, closely related to variations in human movement and behavior.

Although geographical movement is becoming increasing easy to measure through advanced geospatial technologies, a patient's movements and place history have been largely ignored by the medical community. Quantifying and describing human movements provides valuable information necessary to predict disease outbreaks and to evaluate control alternatives to halt epidemics [18–20]. In addition, the ability to apply this knowledge to a variety of diseases creates an opportunity to identify common areas where infection occurs across multiple diseases and to leverage public health programs to target the most important locations that serve as sources for more than one disease. By examining of the role of human movement across different scales, public health communities can use this valuable information on pathogen transmission to increase the effectiveness of disease prevention programs. As transmission rates are reduced through intervention efforts, scientists can expect the importance of heterogeneity in exposure to increase and play an even more important role in pathogen persistence. Therefore, characterization of human and population movements can facilitate not only the elimination of disease, but also help to prevent its return [15].

Key considerations in the analysis of human movement patterns include, but are not limited to: spatial scale, the type and periodicity of movement, and the time period of observation. Spatial scale refers to the geographical extent of the pathogen and the spatial interpretation of the data. The geographical extent of the pathogen and its transmission can be determined by the disease dynamic – i.e., the spread of a pathogen to new geographical areas versus a sustained transmission at a given location. If the transmission is local, then the relevant movements will be those placing the susceptible hosts in high risk locations at times when infection risk is high. Assumptions regarding the importance of movements should be made with care because heterogeneity in exposure can have a dramatic effect on infection risk.

The type of movement refers to what the researcher is aiming to measure. For instance, is the study interested in the sites where individuals spend their time on a regular basis (high spatial and temporal resolution) or when they are traveling outside of their home city? What is the value of travel information (outside of an urban area) that specifies exactly where people go? Are specific routes important, or should only destinations be considered? These specific details will depend on the nature of the questions, systems, and resources and methods involved for measuring movements.

Finally, the time period of observation is concerned with how long to observe individual movements for. The correct answer will depend on the questions being asked and available resources. In the case of dengue, infection can occur up to 2 weeks prior to the manifestation of symptoms. For a retrospective study, 14–15 days would be an appropriate observation period. Conversely, in a prospective study, the length of the observation period will depend on the relative importance of rare movements. Studies of human movements in developed societies reveal markedly regular patterns, especially during the work-week [19, 21, 22]. Additionally, there may be significant instability in movements on weekends or at other times (e.g., vacations). For regular movements during the work week, at least 2 weeks of observation are desired. For more variable movements/times, longer observation periods will be necessary.

8.4 Linking the Global to the Local Through Mobility

Over the last century, human civilizations and urbanization have witnessed a huge increase in mobility and population growth. The combined effect of these factors means that, despite great improvements in hygiene, sanitation and vector control, the containment of disease remains one of the biggest challenges of our modern contemporary society. The importance of increased human mobility in disease transmission cannot be under-stated. On national and global scales, the airline transport network has played a key role in the global dissemination of influenza and SARS [23]. In addition, migrants, tourists, and commercial travelers also have a significant influence on the spread of HIV [24]. On a local scale, human movements in metropolitan areas are frequent and extensive, but are often composed of highly structured commuting patterns between the home and places of employment, education, or business. The ability to quantify human movement and its effects are vital in the ongoing development of strategies to eradicate vector-borne diseases (such as dengue) from urban centers [25].

Using a meta-population analysis where mobile humans connect with static mosquito subpopulations in a very structured pattern, a group of researchers found that, due to frequency dependent biting, infection incidence in the human and mosquito populations is independent of the duration of contact [26]. The researchers hypothesize that, since the biting rate is frequency dependent (and independent of the density of the human population), a mosquito will bite the same number of people per day. In addition, their modeled results indicate that people who travel regularly to areas with large mosquito populations form a high-risk group, and have a relatively high level of infection compared to people that travel regularly to patches with small mosquito populations. Furthermore, extensive variation in human movement patterns causes the number of interactions between human and mosquito populations to increase. More variable human movements increases the likelihood that people will carry the infection from these highly infested areas to mosquito subpopulations where the pathogen has died out. Therefore, a large mosquito population in a frequently visited area may be sufficient to ensure infection is endemic, even if there are relatively few mosquitoes elsewhere. When people do not vary their travel patterns very much and there is no direct connectivity between mosquito populations, the transit corridor can significantly enhance disease persistence by acting as a reservoir and hub. If people vary the areas they visit even occasionally, the effect of the transit corridor is overridden [26, 27].

Mosquito and human movements become even more important as remote rural villages are connected to each other by mass transportation networks. Nearly a century ago, it was observed that people do not develop a disease where it is contracted or even close to that place [28]. Today, widespread mass transportation makes that observation even more relevant. The incidence and persistence of vector-borne diseases on relatively small spatial scales may be strongly influenced by infectious humans who remain mobile because the infection is mild or silent. Increased human movement on a local scale may be a key factor behind increased incidence of vector-borne diseases.

In modern metropolitan areas, daily travel is a common way of life. Distant subpopulations of mosquitoes may be connected by this daily movement. Large, localized mosquito populations in areas that people visit regularly may be both reservoirs and hubs of infection, even if people only pass through those locations briefly. This implies that surveillance with the goal of controlling vector-borne disease may be a much greater challenge than originally anticipated.

Ultimately, successful public health intervention must focus on both hosts and vectors. Large mosquito populations that are also visited by a large fraction of the human population need to be identified. It is vital to employ surveillance strategies that reveal the variability in the distribution of mosquitoes and target areas where the mosquito population is significant and human movement is extensive. Further study of networks formed by human movement in urban areas are need, and cell phone records are one potential source of such detailed information [19].

References

1. Institute of Medicine (2009) Committee on the U.S. Commitment to Global Health. The U.S. commitment to global health: recommendations for the public and private sectors. National Academies Press, Washington, DC
2. Fineberg HV, Hunter DJ (2013) A global view of health – an unfolding series. NEJM 368(1):78–79
3. Bejon P, Williams TN, Liljander A et al (2010) Stable and unstable malaria hotspots in longitudinal cohort studies in Kenya. PLoS Med 7(7):e1000304. doi:10.1371/journal.pmed.1000304
4. Tatem AJ, Smith DL (2010) International population movements and regional Plasmodium falciparum malaria elimination strategies. PNAS 107(27):12222–12227

5. Wesolowski A, Eagle N, Tatem AJ et al (2012) Quantifying the impact of human mobility on malaria. Science 338:267–270
6. Kester JGC (2014) 2013 International tourism results and prospects for 2014. http://dtxtq4w60x-qpw.cloudfront.net/sites/all/files/pdf/unwto_fitur_2014_hq_jk_1pp.pdf. Accessed 14 June 2014
7. International Trade Administration (2013) U.S. travel to international destinations increased three percent in 2012. Department of Commerce, International Trade Administration, Washington, DC. http://travel.trade.gov/outreachpages/download_data_table/2012_Outbound_Analysis.pdf. Accessed 14 June 2014
8. Warren A, Bell M, Budd L (2010) Airports, localities and disease: representations of global travel during the H1N1 pandemic. Health Place 16:727–735
9. Herman JS, Hill DR (2010) Advising the traveler. Medicine 38(1):56–59
10. Wieten RW, Leenstra T, Goorhuis A, van Vugt M, Grobusch MP (2012) Health risks of travelers with medical conditions – a retrospective analysis. J Travel Med 19:104–110. doi:10.1111/j.1708-8305.2011.00594.x
11. Prothero RM (1977) Disease and mobility: a neglected factor in epidemiology. Int J Epidemiol 6:259–267
12. Prothero RM (1963) Population mobility and trypanosomiasis in Africa. Bull World Health Organ 28:615–626
13. Gubler DJ (1997) Dengue and dengue hemorrhagic fever: its history and resurgence as a global public health problem. In: Gubler DJ, Kuno G (eds) Dengue and dengue hemorrhagic fever. CAB International, London, United Kingdom, pp 1–22
14. Messina JP, Moore NJ, DeVisser MH, McCord PF, Walker ED (2012) Climate change and risk projection: dynamic spatial models of tsetse and African Trypanosomiasis in Kenya. Ann Assoc Am Geogr 102(5):1038–1048. doi:10.1080/00045608.2012.671134
15. Stoddard ST, Morrison AC, Vazquez-Prokopec GM, Paz Soldan V, Kochel TJ et al (2009) The role of human movement in the transmission of vector-borne pathogens. PLoS Negl Trop Dis 3(7):e481. doi:10.1371/journal.pntd.0000481
16. Behrens JJW, Moore CG (2013) Using geographic information systems to analyze the distribution and abundance of Aedes aegypti in Africa: the potential role of human travel in determining the intensity of mosquito infestation. Int J Appl Geospatial Res 4(2):9–38
17. Wen T-H, Lin M-H, Fang C-T (2012) Population movement and vector-borne disease transmission: differentiating spatial–temporal diffusion patterns of commuting and non-commuting dengue cases. Ann Assoc Am Geogr 102(5):1026–1037. doi:10.1080/00045608.2012.671130
18. Eubank S, Guclu H, Kumar VSA, Marathe MV, Srinivasan A et al (2004) Modelling disease outbreaks in realistic urban social networks. Nature 429
19. Gonzalez MC, Hidalgo CA, Barabasi A-L (2008) Understanding individual human mobility patterns. Nature 453:779–782
20. Riley S (2007) Large-scale spatial-transmission models of infectious disease. Science 316:1298–1301
21. Hagerstrand T (1970) What about people in spatial science. Reg Sci Assoc 24:7–21
22. Schlich R, Axhausen KW (2003) Habitual travel behaviour: evidence from a six-week travel diary. Transportation 30:13–36
23. Colizza V, Barrat A, Barthelemy M, Vespignani A (2006) The role of the airline transportation network in the prediction and predictability of global epidemics. Proc Natl Acad Sci U S A 103:2015–2020
24. Perrin L, Kaiser L, Yerly S (2003) Travel and the spread of HIV-1 genetic variants. Lancet Infect Dis 3:22–27
25. Gubler DJ (1998) Dengue and dengue hemorrhagic fever. Clin Microbiol Rev 11:480–496
26. Adams B, Kapan DD (2009) Man bites mosquito: understanding the contribution of human movement to vector-borne disease dynamics. PLoS ONE 4(8):e6763. doi:10.1371/journal.pone.000676
27. Longini IM, Koopman JS (1982) Household and community transmission parameters from final distributions of infections in households. Biometrics 38:115–126
28. Conner ME, Monroe WM (1923) Stegomyia indices and their value in yellow fever control. Am J Trop Med Hyg 4:4–19

Chapter 9
Geospatial Medicine

Abstract The emerging field of geospatial medicine is a sub-discipline of medicine that emphasizes the importance of a patient's place history in the diagnosis and treatment of disease. Much like the way in which laboratory test reports flag results that are outside of a test's reference range, GIS technologies can be designed to provide similar warnings, or flags, to help physicians take notice of environmental – or "place" – factors that could be contributing to the patient's symptoms. Geospatial medicine provides physicians with a more precise clinical understanding about where patients live, work, and play, and how a patient's movement and place history can reduce exposure or risks to environmental or social hazards that adversely personal health. The National Children's Study and the Center for Geospatial Medicine are two research programs that are highlighted in this chapter, as these research programs enable clinicians to make connections between multi-dimensional environmental factors, and to provide an integrated therapy that links environmental health information to patient care. In addition, geospatial medicine's focus on personalized care is closely aligned with the goals of accountable care organizations (ACOs), as prescribed by the Centers for Medicare and Medicaid Services (CMS). Finally, as many universities and research institutions have begun developing curricula that teach the relationship between geography and patient care, it is very likely that the next generation of health care providers will be well-trained in the life-saving contributions of geospatial medicine.

Keywords Geospatial medicine • Place history • Geospatial health intelligence • Clinical markers • National Children's Study • Center for Geospatial Medicine • Accountable care organizations

9.1 Introduction

As illustrated throughout this volume, patient health outcomes are shaped by a number of multi-dimensional factors, such as social, environmental, and host-vector influences. Unfortunately, many clinicians and members of the medical community are not readily able to make the connections between these multi-dimensional factors to provide an integrated approach to link environmental health research and patient care. To address this care gap, geographic information systems (GIS) technology and

A.J. Blatt, *Health, Science, and Place*, Geotechnologies and the Environment 12,
DOI 10.1007/978-3-319-12003-4_9

spatial statistics provide physicians with the ability to integrate these multi-dimensional components into a comprehensive model, and facilitate innovative strategies for improving public health and patient care.

The value of spatial analysis is the ability to recognize that virtually all data contain a geographic component that can be tied to a specific location, such as a state, county, ZIP code, or single address, as well as larger geographical features, such as a watershed, air shed, or hospital health system. Geographic analysis allows users to explore and overlay data by location, and to reveal hidden spatial and temporal trends that are not readily apparent in traditional spreadsheet and statistical packages. In addition, GIS technology allows for the construction of spatial and temporal data architectures that can then be analyzed with either spatial or aspatial statistics.

The emerging field of "geospatial medicine" introduces a new type of clinically relevant information that leverages national spatial data infrastructures to improve patient care. By linking a patient's own health status to certain relevant spatial and temporal factors, medical professionals can use a patient's "place history" to improve the quality of health care services and delivery. Geospatial medicine is a new sub-discipline of medicine, and provides medical professionals with a more precise clinical understanding about where patients live, work, and play, and how a patient's movement and place history can reduce exposure or risks to environmental or social hazards that adversely personal health [1].

The concept of "geomedicine" was first published in 1945, in a correspondence from the *British Medical Journal* that described a Haifa hospital medical superintendent's concern that Europeans migrating to Palestine "without the comfort of home and family life… are apt to develop nostalgic reactions and suffer from neuroses." In his letter, Dr. Joseph Seide suggested that "a central Academy of Geomedicine… for studying the problems of settlement of the white race in colonial territories of the Empire and the Dominions situated in hot zones… [is] certainly vital, not only for the British Empire, but for other countries – including mine – which are dealing with immigrants of European origin…" [2]. In 1986 Professor Jul Lag, of the Agricultural University of Norway, published a letter in *Ambio*, and describes geomedicine as a new discipline "that analyzes the impact of ordinary environmental factors on the geographical distribution of human and animal health problems" [3]. Indeed, the history of geomedicine can be traced back to mid-1850s, when a physician, Dr. John Snow, used pinpoint mapping to prove his theory that cholera is a water transmitted disease. By spatially superimposing cholera deaths with London's public water supplies, he identified the Broad Street pump as the probable source of the cholera outburst (Fig. 9.1). The subsequent removal of these water pump by the authorities significantly reduced the number of cholera deaths in London [4].

As the previous examples have illustrated, a patient's "place history" is a key clinical marker for environmental and social exposures, such as air, water, ground, food, culture, demographic characteristics, and service delivery capacities. A patient's address history is a clinically relevant piece of health information, in understanding a patient's environmental, social and cultural history. Currently, most of this information is not immediately accessible by a clinician at the time of a patient encounter

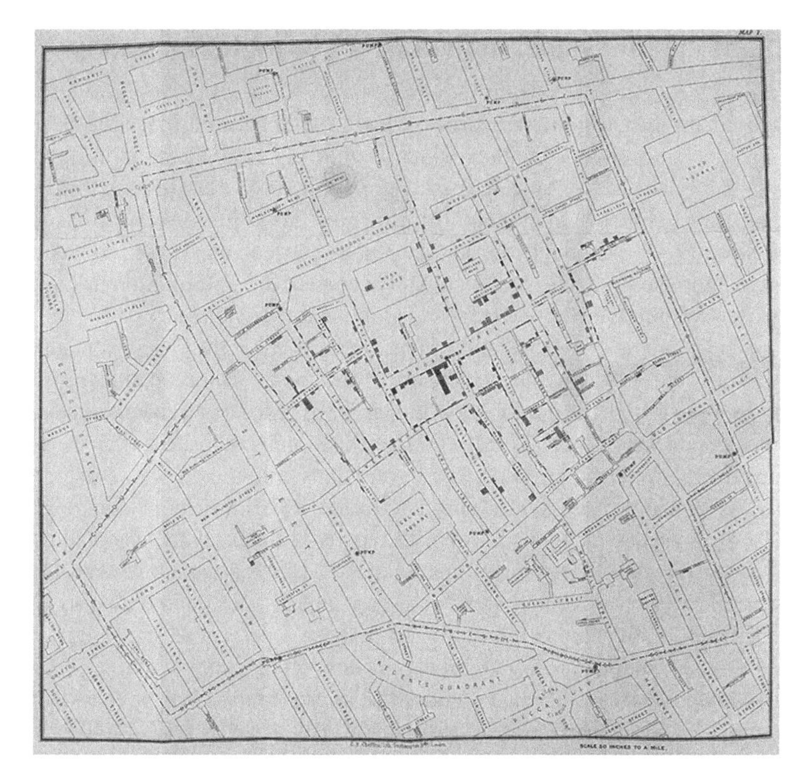

Fig. 9.1 Dr. John Snow's famous cholera map. The Broad Street pump is indicated by a smudged area, center-left. The stippled line demarcates equal walking distances between the pump in Broad Street and the nearest rival pump at every point (From: The British Library Board, Available at: http://www.bl.uk/learning/images/makeanimpact/publichealth/large12735.html, Accessed 28 June 2014)

and, unfortunately, is not a part of a physician's comprehensive assessment of a patient's medical condition. GIS technology offers a solution to this problem in that it delivers real-time geospatial medical information to the patient examining room and adds a critical piece of medical intelligence that allows clinicians to ensure the health and safety of their patients. Simply put, in the context of healthcare, GIS technology adds geospatial health intelligence to a patient's medical history [1].

This chapter will describe the importance of geospatial medicine in health care delivery and access by discussing how population health can be linked to patient health through knowledge of the patient's movements and place history. Then, it will illustrate how two major research programs within the United States are molding the future of this new sub-discipline of medicine. Finally, this chapter will conclude with a number of future directions for geospatial medicine that promises to bring geospatial health intelligence to every patient examining room in the country.

9.2 Linking the Global to the Local Through Geography

Throughout the last decade, researchers have demonstrated that area-based, or geographic, socioeconomic characteristics provide a good indication of individual-level characteristics [5, 6]. Indeed, these geographic measures are used to illustrate and monitor socioeconomic inequalities when individual-level monitoring is not possible [7–9]. Although the use of geographic, or area-based, measures to understand population health has been well accepted, this information has not been widely used to improve health care at the individual patient level.

In a novel example of geospatial medicine, researchers at the Cincinnati Children's Hospital Medicine center investigated the use of area-based socioeconomic measures to identify children at high risk for further asthma-related utilization (readmission or return to the emergency department). They found that children living in areas characterized by extreme poverty, lower home value, and fewer adults with a high school degree were strongly associated with asthma-related re-utilizations [10]. The population-level data were also strongly associated with caregiver reports of financial hardship and psychological distress, both of which have been linked to adverse asthma outcomes [11–13]. This study demonstrated that an area-based geographic index can be used to identify children at high risk for further morbidity and potential intervention strategies.

Indeed, a key component of geospatial medicine is the ability to utilize geographic information about environmental conditions to improve individual-level patient care. An example of this is the ability of asthma patients to use a rescue inhaler that is equipped with a global positioning system (GPS) sensor. Such a device maps the patient's location every time a puff is taken and sends that information back to the physician, who can then pinpoint the environmental causes and design the appropriate treatment plans. When a prototype of a GPS-enabled inhaler was first developed, epidemiologist David Van Sickle hoped that it would provide physicians valuable information about when and where asthma attacks occur. Initially funded by the Centers for Disease Control and Prevention (CDC) in 2006, the prototype has led the inventor to establish a company, now called Propeller Health, which has continued to fine-tune the digital sensor for asthma and chronic obstructive pulmonary disease (COPD). As of this writing, the latest version of the sensor can be attached to any inhaler, and is equipped with a smaller, Bluetooth-based device that sends usage information to a Web portal that can display when and where patients have used their inhalers [14, 15].

Other similar geomedicine technologies under development include Asthma Trigger, a wireless asthma sensor developed by AT&T Labs that scans the ambient air for compounds which may cause asthma symptoms, and alert patients via their mobile devices. The sensor feeds air quality data through AT&T's healthcare information data exchange platform and allows patients to receive alerts on their mobile devices. In addition, researchers from Saint Louis University and Harvard University have received funding to collaborate on an asthma alert messaging system that will use Google Maps software to send alerts to users when outdoor conditions are such that they could potentially cause an asthma attack [15].

Just as knowledge of a patient's DNA can help physicians deliver personalized medicine, geospatial medicine uses a physician's knowledge of a patient's environmental

DNA (or place history) to improve health care services and delivery. The ability to integrate a patient's place history into existing electronic medical and personal health records is a key role of GIS technology. By adding place history to a patient's medical history, geospatial medicine provides a more accurate picture of a patient's health that is also important for other aspects of medical care, such as follow-up care, reminder notices, and health alerts [16].

Much like the way that a person's family medical history helps physicians detect certain predispositions to diseases, knowledge of a patient's place history provides the context within which clinicians can assess environmental factors and make judgments about the health and treatment of the patient. Moreover, just as laboratory test reports flag results that are "out-of-range," GIS technologies can be designed to provide similar warnings, or flags, to help physicians take notice of environmental – or "place" – factors that could be contributing to the patient's symptoms. In this way, GIS plays a major role in helping physicians guide patients in making lifestyle changes that could impact their health in a significant manner.

9.3 Research Programs in Geospatial Medicine

The promise of geospatial medicine is that it can provide a large body of clinically relevant knowledge that explains many of the confounding environmental problems that confront contemporary lifestyles. This is important because geospatial health intelligence enables clinicians to extend their diagnostic, treatment, and preventative reach beyond what is available to them today. GIS technology is a valuable asset that can be leveraged successfully to benefit the many healthcare changes that Americans are experiencing today – one which seeks to improve personal health status and affordably deliver our society's necessary health services.

The two research programs highlighted below contribute the clinician's ability to make connections between multi-dimensional environmental factors, so that he/she can provide an integrated therapy that links environmental health research to patient care. These research programs represent significant financial investments in the health and well-being of several generations of Americans. The Center for Geospatial Medicine and the National Children Study both recognize the clinical value of understanding the genetic, environmental and cultural influences on human health and development. The use of GIS technology and the importance of a patient's place history are important components of these studies. The descriptions below focus on the contributions and successes of each research program.

9.3.1 Center for Geospatial Medicine

The Center for Geospatial Medicine (CGM) is a research center housed in the University of Michigan and Duke University Medical Center that focuses on developing geospatial informatics tools for analyzing the genetic, environmental, and psychosocial pathways that jointly shape health and well-being. In 2011, CGM received a $6.2 million grant from the Bristol-Myers Squibb Foundation as part of its national diabetes initiative, Together on Diabetes. The project's mission is to improve health outcomes and the quality of life for people living with Type 2 diabetes in Durham County, North Carolina, where nearly 10 % of its residents have already been diagnosed with diabetes [17]. Nationally, diabetes affects approximately 29.1 million Americans, or 9.3 % of the country's population [18].

The Center for Geospatial Medicine specializes in the use of geospatial medicine to enable health officials integrate health care from the household level to the specialty clinics, in order to provide patients with the tools to improve their health outcomes. CGM uses GIS technology to enable researchers to visualize the complex relationships among the locations of diabetes patients, patterns of health care, and available social resources. Using a systematic, spatially based methodology, researchers can identify gaps in access to care and self-management resources; help patients connect with the community assets; and identify interventions that can result in better health outcomes – both for the individual and the neighborhood as a whole. In addition, the geospatial data can be used to create a continuous feedback loop for improving the quality of project efforts [17].

Recognizing the important clinical impact of this study, the Department of Health and Human Services also presented the Center for Geospatial Medicine with a $9.7 million Health Care Innovation award, in 2012, to reduce death and disability from Type 2 diabetes among at-risk populations in four underserved Southeastern counties (Durham County, NC; Cabarrus County, NC; Quitman County, MS; and Mingo County, WV). The project plans to provide patient-centered coordinated care to improve individual-level patient health outcomes and lower health care costs by over $20 million through the reduction of hospital and emergency room admissions and the use of preventive care to curb the need for amputations, dialysis, and cardiac procedures [19].

9.3.2 National Children's Study

The over-arching goal of the National Children's Study (NCS) is to improve the health and well-being of children by understanding the effects of environment and genetics on the growth, development, health of children. The National Children's Study recognizes the importance of place history and allows researchers to follow 100,000 children for 21 years – from before birth until age 21 – to determine how genetic and environmental factors such as exposure to natural and manufactured products, noise and stress levels, air, water, and soil affect children during different phases of their lives. In addition – although the term "geomedicine" is not explicitly used in the study guidelines – the NCS broadly defines the environment to include air, water, diet, sound, family dynamics, community and cultural influences [20].

The NCS is dedicated to promoting the well-being of several generations of Americans by being the largest and most comprehensive children's health research in the United States. Since personal health begins before birth and continues throughout person's life, understanding how environmental exposures interact with a child's genetic background will address a significant knowledge gap in the improvement of patient care. Several design components of the NCS are worthy to note. First, the NCS includes the largest representative sample of children from different socioeconomic, ethnic and racial backgrounds. Second, the study monitor mothers during pregnancy to determine how prenatal factors may affect the health of a child. Finally, by examining how a wide range of environmental exposures impact children's health, the NCS recognizes that the environment may have a different impact on children than adults [21].

Funding for the NCS is substantial. In 2007 and 2008, the United States Congress appropriated funds for the first phase of implementation of the National Children's Study. These funds enabled the formation of 36 Study Centers. In fiscal year 2009, funding for the National Children's Study was $179.7 million; in fiscal year 2010, funding was $193.9 million; in fiscal year 2011, funding was $191.1 million; and enacted funding for fiscal year 2012 was $193.1 million [22].

This ambitious study is spearheaded by the Eunice Kennedy Shriver National Institute of Child Health and Human Development of the National Institutes of Health (NIH), in collaboration with a consortium of federal government partners. The federal government study partners include the National Institute of Environmental Health Sciences of the NIH, the Centers for Disease Control and Prevention, and the Environmental Protection Agency (Fig. 9.2).

9.4 The Promising Future of Geospatial Medicine

Through a number of different avenues, medicine is making major advances towards painting a more holistic picture of patient health. One factor that can be included in a patient's medical history is a history of where that patient has lived. This information can be combined with other commonly collected genetic and lifestyle information to offer a better picture of health and more accurate indications of future disease or areas of concern. Moreover, this information enables patients to create more awareness about their own health. A limitation, however, of factoring in geography to an overall health picture is that, like other health factors, individual outcomes will vary. Therefore, it is very important to recognize that geography is just one factor in the total overall picture of patient health. For instance, just because a patient grew up in a heavily industrial area does not necessarily imply that the patient will suffer from cardiovascular disease. The benefits of applying geography to patient care is that it does provide a very useful indicator in painting a more holistic picture of a patient's health.

Clinicians and physicians should advocate for the inclusion of geography in a patient's electronic medical record (EMR). Currently, doctors ask their patients for information about their genetics (e.g., does anyone in your family have a history of breast cancer…) and lifestyle choices (e.g., do you smoke…). Knowledge of a

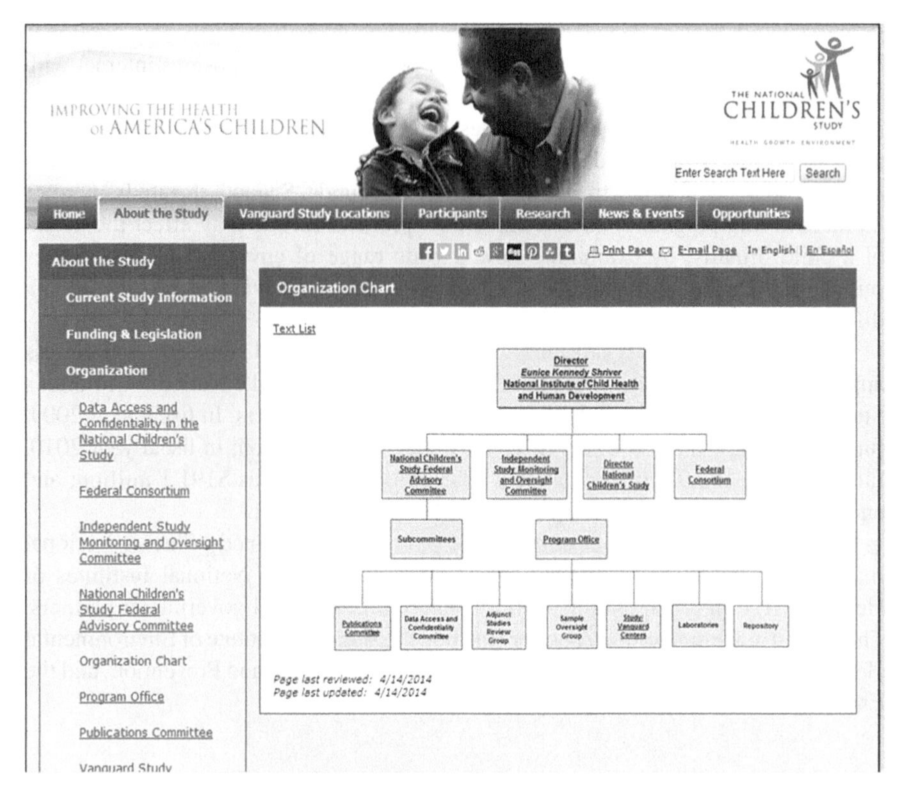

Fig. 9.2 The National Children's Study is led by the National Institutes of Health (NIH) (Source: http://www.nationalchildrensstudy.gov/about/organization/organization_chart/Pages/default.aspx, Accessed 30 June 2014)

patient's place history can help physicians understand the patient's overall health, which leads to an enhanced ability to prescribe preventative care.

Geospatial medicine's focus on personalized care is closely aligned with the objectives of accountable care organizations (ACOs), as prescribed by the Centers for Medicare and Medicaid Services (CMS). Their common goal is to provide cost-effective preventative care and a highly satisfactory health care experience for each individual patient. However, privacy concerns over patient confidentiality and geographic profiling may slow the adoption of geospatial medicine in medical practices and hospitals. For instance, potential data breaches would now compromise very detailed information surrounding a patient's daily routines. Furthermore, health privacy experts are wary of the pitfalls of geographic profiling by insurance companies which may lead to a denial of insurance coverage to patients living in areas with multiple environmental hazards [23].

While geospatial medicine is still in its infancy, its potential to improve patient care is enormous. A key step in the acceptance of geospatial medicine by physicians will be the accessibility and actionability of research in the fields of medical geography and spatial epidemiology. The ability of medical libraries and health research organizations to make electronic geospatial databases accessible over the Internet is crucial to the adoption of geospatial medicine in medical practice [24, 25]. In addition, as many universities and research institutions have begun developing curricula and programs that teach the relationship between geography and patient care, it is very likely that the next generation of health care providers will be well-versed in the life-saving contributions of geospatial medicine [26].

References

1. Davenhall B (2012) Geomedicine: geography and personal health. Esri, Redlands. http://www.esri.com/library/ebooks/geomedicine.pdf. Accessed 29 June 2014
2. Seide J (1945) Geomedicine. Br Med J 2(4433):896
3. Lag J (1968) A place for geomedicine. Ambio 15(5):313
4. Snow J (1855) Dr. Snow's report. In: Report on the Cholera Outbreak in the Parish of St. James, Westminster, during the autumn of 1854, Churchill, London, pp 97–120. http://johnsnow.matrix.msu.edu/work.php?id=15-78-55. Accessed 28 June 2014
5. Subramanian SV, Chen JT, Rehkopf DH, Waterman PD, Krieger N (2006) Comparing individual- and area-based socioeconomic measures for the surveillance of health disparities: a multilevel analysis of Massachusetts births, 1989–1991. Am J Epidemiol 164(9):823–834
6. Rehkopf DH, Haughton LT, Chen JT, Waterman PD, Subramanian SV, Krieger N (2006) Monitoring socioeconomic disparities in death: comparing individual-level education and area-based socioeconomic measures. Am J Public Health 96(12):2135–2138
7. Krieger N, Chen JT, Waterman PD, Rehkopf DH, Subramanian SV (2005) Painting a truer picture of US socioeconomic and racial/ethnic health inequalities: the Public Health Disparities Geocoding Project. Am J Public Health 95(2):312–323
8. Krieger N, Chen JT, Waterman PD, Rehkopf DH, Subramanian SV (2003) Race/ethnicity, gender, and monitoring socioeconomic gradients in health: a comparison of area-based socioeconomic measures – the Public Health Disparities Geocoding Project. Am J Public Health 93(10):1655–1671
9. Eibner C, Sturm R (2006) US-based indices of area-level deprivation: results from HealthCare for Communities. Soc Sci Med 62(2):348–359
10. Beck AF, Simmons JM, Huang B, Kahn RS (2012) Geomedicine: area-based socioeconomic measures for assessing risk of hospital re-utilization among children admitted for asthma. Am J Public Health 102:2308–2314. doi:10.2105/AJPH.2012.300806
11. Williams DR, Sternthal M, Wright RJ (2009) Social determinants: taking the social context of asthma seriously. Pediatrics 123(suppl 3):S174–S184
12. Wright RJ (2011) Epidemiology of stress and asthma: from constricting communities and fragile families to epigenetics. Immunol Allergy Clin North Am 31(1):19–39
13. Beck AF, Huang B, Simmons JM, Moncrief T et al (2014) Role of financial and social hardships in asthma racial disparities. Pediatrics 133(3):431–439. doi:10.1542/peds.2013-243
14. Propeller Health. http://propellerhealth.com/. Accessed 29 June 2014
15. Slabodkin G (2013) Growing field of 'geomedicine' offers great promise for asthmatics. http://www.fiercemobilehealthcare.com/story/growing-field-geomedicine-offers-great-promise-asthmatics/2013-02-18. Accessed 29 June 2014
16. Davenhall B (2013) A need to know: adding DNA and geomedicine data to patient records. http://www.xconomy.com/san-diego/2013/12/24/need-know-adding-dna-geomedicine-data-patient-records/?single_page=true. Accessed 29 June 2014
17. Morrill K (2011) U-M, Duke Team to improve diabetes outcomes. http://www.snre.umich.edu/highlights/2011-11-15/um_duke_team_to_improve_diabetes_outcomes. Accessed 30 June 2014

18. Centers for Disease Control and Prevention (2014) National diabetes statistics report: estimates of diabetes and its burden in the United States, 2014. U.S. Department of Health and Human Services, Atlanta. http://www.cdc.gov/diabetes/pubs/statsreport14/national-diabetes-report-web.pdf. Accessed 30 June 2014
19. Managed Care Law Weekly (2012) Diabetes; Southeast program to fight diabetes awarded nearly $10 million by HHS. NewsRx, Atlanta, pp 3548–3549
20. National Institutes of Health, Eunice Kennedy Shriver Institute of Child Health and Human Development (2014) National children's study. http://www.nichd.nih.gov/about/org/od/ncs/Pages/index.aspx. Accessed 30 June 2014
21. Guttmacher AE, Hirschfeld S, Collins FX (2013) The national Children's study – a proposed plan. NEJM 369(20):1873–1875
22. National Children's Study (2014) Congressional history. http://www.nationalchildrensstudy.gov/about/funding/Pages/congressionalhistory.aspx. Accessed 30 June 2014
23. Bresnick J (2013) Healthcare foursquare? Geomedicine tracks disease triggers. EHR Intelligence. http://ehrintelligence.com/2013/02/07/healthcare-foursquare-geomedicine-tracks-disease-triggers/. Accessed 30 June 2014
24. Cromley E (2011) The role of the map and geographic information library in medical geographic research. J Map Geogr Libr 7(1):13–35
25. Blatt A (2011) Maps, geography libraries, and health outcomes: gazing into the future of medical geography. J Map Geogr Libr 7(1):2–12
26. Hayashi S, Bazemore A, McIntyre J (2011) Transforming community health and primary care education using clinical and administrative data and geographic information systems. J Map Geogr Libr 7(1):61–70

Chapter 10
A New Model

Abstract This final chapter summarizes the purposes and goals of the Patient Protection and Affordable Care Act (PPACA), and describes how it is organized to expand health care coverage and encourage disease prevention. Having a patient's medical information (such as notes from health care visits, laboratory and radiology test results, prescribed medications, and health insurance) stored in an electronic health record (EHR), and sharing that information with different clinicians involved in a patient's care, is the promise of health information exchanges (HIEs). Drawing upon knowledge from different disciplines, a new model of medical geography is described, one that re-contextualizes the discipline against a history of national health care reform, and advocates for a more engaged model of patient health care that utilizes geospatial health intelligence in the examining room.

Keywords The Patient Protection and Affordable Care Act • Electronic health records • Health information exchanges • Medical geography • Health care reform • Patient engagement

10.1 Introduction

The purpose of this book is to paint a new portrait of medical geography, one that frames it in a more centralized role on the national stage of the U.S. health care reform. By presenting a new model of patient care that engages the patient's geographic and medical history, this book draws upon knowledge from seemingly disparate disciplines – such as environmental geography, transportation geography and geospatial medicine – to weave a tapestry that connects population health to patient care, against the backdrop of health care reform in the United States. Seizing on the current attention on health information exchange (HIEs), this patient-centric view takes advantage of the current advances in geospatial data and technologies to advocate the use of place histories as a key clinical marker in the diagnosis, prognosis, and treatment of disease.

In this final chapter, we return to the beginning of this volume and revisit President Obama's goal for the Patient Protection and Affordable Care Act (PPACA), which is to espouse "the core principle that everybody should have some basic security when it comes to their health care" [1]. To address certain notable deficiencies

A.J. Blatt, *Health, Science, and Place*, Geotechnologies and the Environment 12,
DOI 10.1007/978-3-319-12003-4_10

in the quality of health care in the United States, the PPACA established a National Strategy for Quality Improvement in Health Care [2]. This national strategy has a three-part goal: to improve the overall quality of health care by emphasizing patient-centered, reliable, accessible, and safe health care; to improve population health using proven interventions that address the behavioral, social and environmental determinant of health; and to reduce the cost of health care for patients, providers, and governments.

In addition to placing a strong emphasis on patient-centric health care, the PPACA established the National Prevention, Health Promotion and Public Health Council, which created and implemented the National Prevention Strategy (NPS). The NPS identified four strategic directions and seven priorities that have the greatest effect on population health in the United States (Fig. 10.1) [3]. There is a focus on increasing American's access to preventive services because, despite evidence that effective clinical preventive services can reduce premature disease and deaths, tens of millions of Americans are not using these services. To address this problem, the PPACA mandated that new private insurance plans and states with expanded Medicaid programs provide a set of clinical preventive services that are recommended

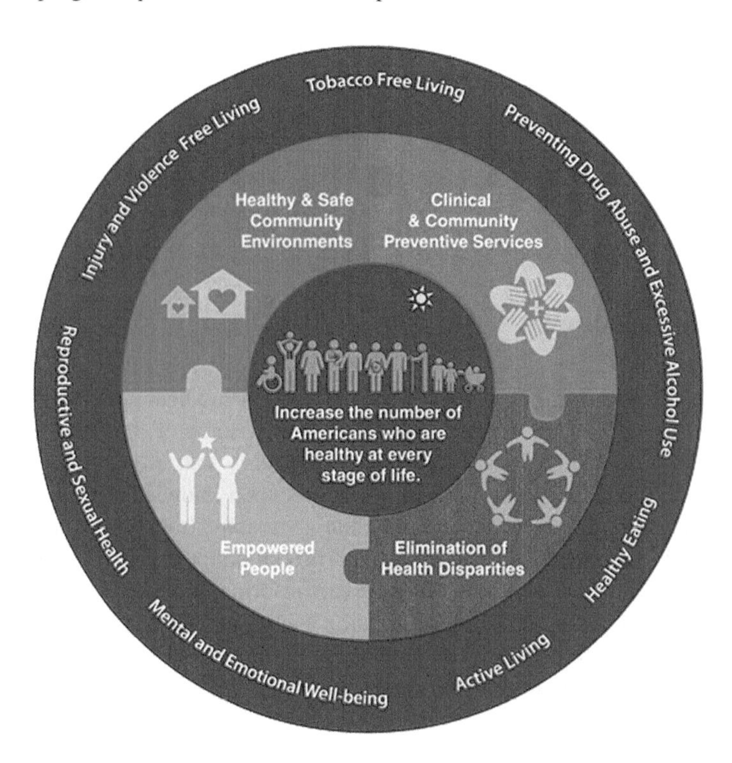

Fig. 10.1 Strategic directions and priorities of the National Prevention Strategy, which "envisions a prevention-oriented society where all sectors recognize the value of health for individuals, families, and society and work together to achieve better health for Americans." (Source: National Prevention Council [3]. National prevention strategy: America's plan for better health and wellness. http://www.surgeongeneral.gov/initiatives/prevention/strategy/, Accessed 4 July 2014)

by the United States Preventive Services Task Force (USPSTF) and the Advisory Committee on Immunization Practices (ACIP), as well as a set of preventive services for women recommended by the Institute of Medicine [4–6].

Furthermore, the PPACA established a Prevention and Public Health Fund to "provide for expanded and sustained national investment in prevention and public health programs to improve health and help restrain the rate of growth in private and public health sector health care costs" [7]. The Prevention and Public Health Fund supports a number of public health and disease prevention activities, such as programs to improve the capacity of state and local public health departments to detect and control disease outbreaks.

As the above programs indicate, the intent of the PPACA is to encourage better ways of organizing health care services and delivery, and the use of evidence-based medicine to improve the health of patients and communities. It is anticipated that this new collaboration will ultimately focus more attention on the prevention of disease (through the detection, control and prevention of disease outbreaks). Throughout this discussion, it has become evident that medical geographers have a crucial role in the implementation of the PPACA – especially in connecting public health disease surveillance to enhanced patient care outcomes. Through the use of geospatial data and technologies, medical geographers are in a unique role of possessing the domain-specific knowledge to add to our current understanding of the cultural, social, and environmental determinants of health. Indeed, as the Center of Geospatial Medicine and the research collaboration between the National Institutes of Health and the Association of American Geographers indicate, medical geographers have an exciting opportunity to address the challenges of providing more efficient and higher quality health care through the use of geospatial data infrastructures, volunteer geographic information, and geospatial data mining algorithms.

In summary, the story of health care reform in the United States is just beginning to unfold and its effects will be felt over many years to come. The promise of geospatial medicine is to translate some of the PPACA's strategic goals into improving patient health care access and delivery. In order for these goals to be realized, clinicians and physicians need to advocate for the inclusion of geography in a patient's electronic medical record (EMR). Medical libraries and health research organizations need to make electronic geospatial databases easily and intuitively accessible over the Internet; in addition, more universities and research institutions need to deliver curricula and programs that teach the relationship between geography and patient care.

10.2 Importance of Geography in Health Care Reform

Public health disease surveillance provides vital information for health officials in several ways – such as the identification and treatment of populations most affected by a disease outbreak, the development of solutions and control measures, and the monitoring of critical intervention efforts. Traditional sources of public health disease surveillance data can be augmented by non-traditional uses of data from other systems (such as volunteered geographic information) and novel uses of new techniques (such as collaborative mapping). In addition, certain health information technologies can substantially increase the efficiency and timeliness of public health surveillance efforts. For instance, meaningful use standards – such as those as implemented from the American Reinvestment and Recovery Act of 2009 – can accelerate the reporting of disease cases to state disease registries. The efficient execution of these registries can expand the current understanding of timeliness of care, effective treatments, and disparities outcomes [8]. Moreover, the inclusion of patient residence histories in HIEs can provide a key clinical marker in the diagnosis, prognosis, and treatment of chronic and infectious diseases.

Geography provides a medical intelligence that leverages one major byproduct of health care reform – national geospatial data infrastructures – to benefit personal human health. As explained in Chap. 7, geospatial data infrastructures are collections of multiple large datasets that are designed to manage and relate information from different fields. For instance, a health-enabled geospatial data infrastructure can be designed to contain and relate geographic, social, and environmental information – and how they are connected to a patient's personal health status. The benefit of making this information easily accessible to clinicians is that this insight can improve patient care by providing a more precise clinical understanding of how personal health is affected by where a patient lives, works, and plays. Although there is currently very little clinically relevant geographic information available to a physician at the time of a medical diagnostic encounter, geographic information systems (GIS) technology promises to deliver vital information concerning a patient's potential environmental exposures and assist a physician in his clinical diagnosis, prognosis, and treatment.

10.3 Engaging a New Model of Patient Care

This section describes how a new model of medical geography can contribute to the quality of health care at five geographic scales: international, national, regional, community, and individual provider. Where possible, actual use cases of quality improvements in health outcomes are provided, in the context of national health reform in the United States and abroad.

10.3.1 International Organizations Use GIS to Eradicate Polio in India

Polio is a crippling infectious disease caused by a virus that spreads from person to person. The virus invades the brain and spinal cord, causing paralysis and sometimes death. Because there is no cure for polio, vaccination is the best and only way to stop the disease from spreading. In the late 1940s to the early 1950s, polio crippled around 35,000 people annually in the United States, making it one of the most feared diseases of the twentieth century. However, due to highly effective

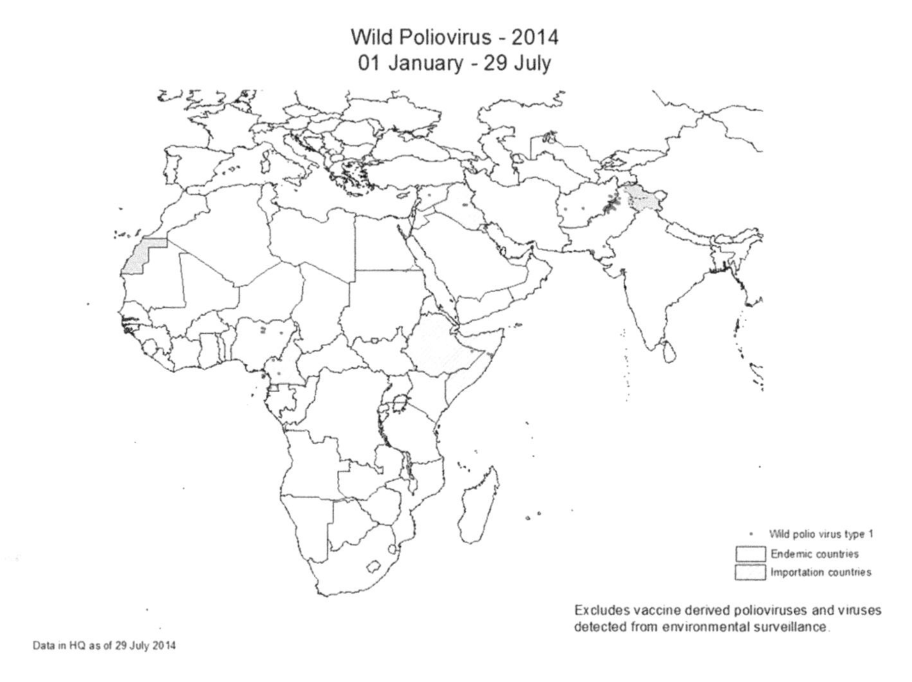

Fig. 10.2 Distribution of the wild poliovirus, as of July 29, 2014 (Source: World Health Organization. Global Polio Eradication Initiative. http://www.polioeradication.org/Dataandmonitoring/Poliothisweek/Poliocasesworldwide.aspx, Accessed 2 Aug 2014)

vaccination campaigns, the United States became polio-free in 1979. Unfortunately, the spread of polio has never stopped in Afghanistan, Nigeria and Pakistan, enabling the reintroduction of the poliovirus to several other parts of the world, such as Syria and Cameroon (Fig. 10.2) [9, 10].

With substantial support from the Bill and Melinda Gates Foundation today, the Global Polio Eradication Initiative (GPEI) was established in 1988 as a public-private partnership comprised of 194 national governments, the World Health Organization, Rotary International, the United States Centers for Disease Control and Prevention, and the United Nations International Children's Emergency Fund (UNICEF) to eradicate polio world-wide. A goal of the GPEI polio eradication strategy is to administer multiple doses of oral polio vaccine (OPV) to all children in their first year of life, through both national and local vaccination campaigns, especially in the highest risk countries. These efforts involve door-to-door immunization in areas where the poliovirus is known or suspected to be circulating, as well as in areas at risk with high population density and mobility, poor sanitation, and low routine immunization coverage. The eradication of polio in India is an excellent example of how geographic information systems (GIS) technology can be leveraged to prevent the spread of disease and assist in nation-wide eradication efforts [11].

Prior to 2014 – when the country was declared polio-free – India was considered one of the most difficult places to eradicate polio (Fig. 10.3). A high population density, high rates of migration, poor sanitation, high birth rates, and low rates of

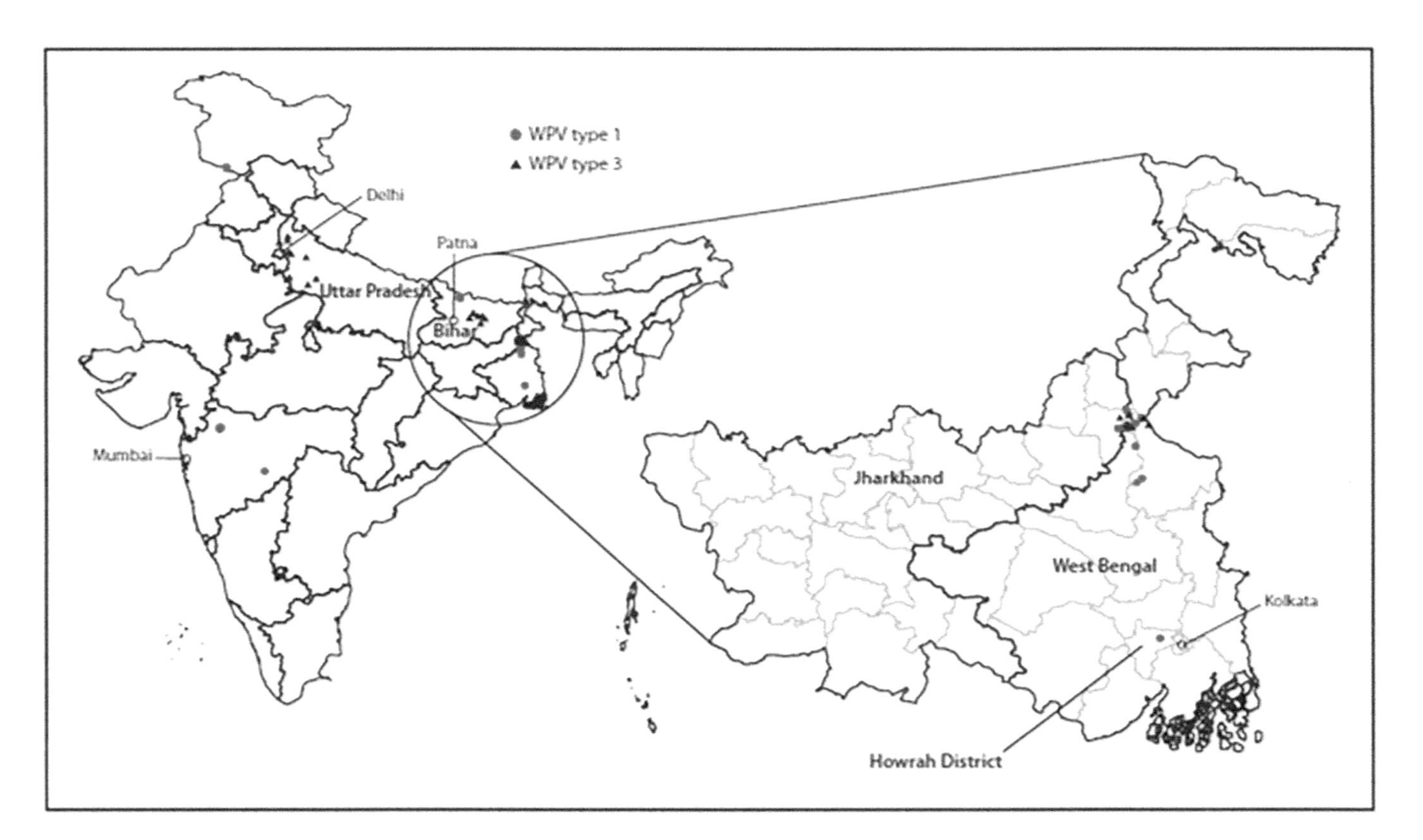

Fig. 10.3 Wild poliovirus (WPV) cases (N = 43), by type in India and selected states during 2010 and 2011. During 2010, a total of 42 WPV cases (18 WPV1 and 24 WPV3) were reported in India in 17 districts in seven states (Source: CDC. Progress toward poliomyelitis eradication – India, January 2010–September 2011. MMWR 2011; 60: 1482–6)

routine immunization were major obstacles to the eradication effort. The success of a vaccination strategy is dependent on microplans, which are detailed logistical blueprints on the size and location of target populations in a given area. These microplans effectively guide the planning and implementation of vaccination campaigns, routine immunization outreach, and surveillance for polio cases by using target population numbers to determine the amount of vaccine required, the number of health care workers and supervisors to deliver the vaccine, and the cost of transportation to get the vaccine and health care workers where they need to go.

Local knowledge of the terrain, seasonal weather patterns and geographic boundaries of an area are critical in creating an effective vaccination strategy. In many countries – like India – this knowledge is out-of-date and imprecise at subnational levels. Especially challenging are rural areas, where household compounds may be several kilometers apart, with no road connecting them. In areas where nomadic pastoralist lifestyles prevail, the situation becomes even more complex, as vaccination team needs to consider certain tribe-specific migration patterns, such as season, type of livestock, family versus clan movement, the role of children in animal herding, and the location of livestock markets (Fig. 10.4).

Therefore, the lack of access to accurate maps led to ambiguity and confusion as to the jurisdiction of smaller, less densely populated settlements. Using GIS and global positioning systems (GPS) technology, members of polio vaccination programs asked local inhabitants to draw the key features of unmapped areas and compared these drawings with published maps and satellite imagery to identify any areas that may have been overlooked in the initial campaigns. In addition to finding unvaccinated children, the GPEI uses GIS to assess the effectiveness of its vaccine drives and predict where reinfection is likely to occur in the future [12].

Fig. 10.4 Intensive mapping campaigns to locate migrant populations and incorporate them into immunization plans in India resulted in the eradication of polio in that country in 2014 (Source: World Health Organization. 2014. http://www.who.int/features/2014/polio-programme/en/, Accessed 12 Aug 2014)

10.3.2 Great Britain's National Health Service Uses Area-Based Analysis to Allocate Funds for Mental Health Services

It is widely recognized that strategic geographic service planning can be used to effectively control rising health care costs in health service systems. Such planning approaches typically monitor variations in the use of services from one local area to another and investigate the factors that seem to explain the variation. For instance, geographical variability in the relative rate of psychiatric service use is important for the planning of mental health services because it can affect the pressures of demand placed on local services and the types of services configurations that most effectively meet those demand.

In 1993, a study was commissioned by National Health Service to analyze variations in hospital admission rates to psychiatric units in from 1990 to 1992, in small areas (known as "synthetic wards," comprising approximately 10,000 people) in Great Britain. The multilevel regression models revealed that the population variables with the strongest positive correlations with admission rates were those that were characteristic of poor urban areas: the proportion of single parent households, proportion of dependents with no caregiver in the household, proportion of people originating from New Commonwealth countries, proportion of elderly people living alone, measures of premature mortality, and relative numbers who were permanently sick and unable to work [13, 14]. The findings of the study were accepted by the Department of Health and the "York Index" used between 1995 and 2003, as the basis for geographically allocating distributing financial resources for National Health Service psychiatric inpatient care. The index explained variations in psychiatric inpatient use among small areas, with reference to these population factors in order to ensure that resources for local psychiatric services are distributed in proportion to local variations in need and demand [15].

Over the years, the York Index has been modified and replaced by the current resource allocation model for mental health services, known as person-based resource allocation for mental health (PRAMH) model. Like the York Index, the PRAMH model incorporates several geographic variables, such as the existence of a nearby mental health provider and distance to the nearest mental health team base [16].

10.3.3 Geospatial Intelligence Highlights Gaps in Patient Care in U.S. Regional Hospital Systems

In the United States, hospital readmissions are considered adverse patient outcomes that highlight important gaps in the quality of patient care provided. Many patients who are discharged from hospitals experience poor clinical care and fragmented clinical communications. Research has shown that, of the nine million hospitalizations of Medicare patients per year, almost one in five patients are readmitted within a month of discharge and many more return to the emergency room [17]. Although some of these readmissions are necessary to complete the episode of care, many readmissions are caused by inadequate discharge planning, poor care coordination between hospital and community clinicians, and the lack of effective longitudinal community-based care. The additional hospital stay signals that many patients get sicker after their initial discharge, leading to additional tests and treatments, and higher health care costs.

Patients with similar illnesses have very different reasons of hospital readmission depending on where they live. Many patients are readmitted because they live in an area where the hospital is used more frequently as a site of care for illness, leading to both higher initial admissions and higher readmissions. According to a study from the Dartmouth Atlas Project, a Medicare patient's chances of being readmitted to a hospital depends largely on where the patient lives and the hospitals from which the patient was discharged [18]. The study revealed marked variations in the percent of patients readmitted to the hospital within 30 days of an initial medical discharge.

Figure 10.5 shows the extent of the spatial variations for the medical discharges discussed in the Dartmouth report: of the 306 hospital referral regions (HRRs) in the United States, 30-day readmission rates following medical discharge ranged from

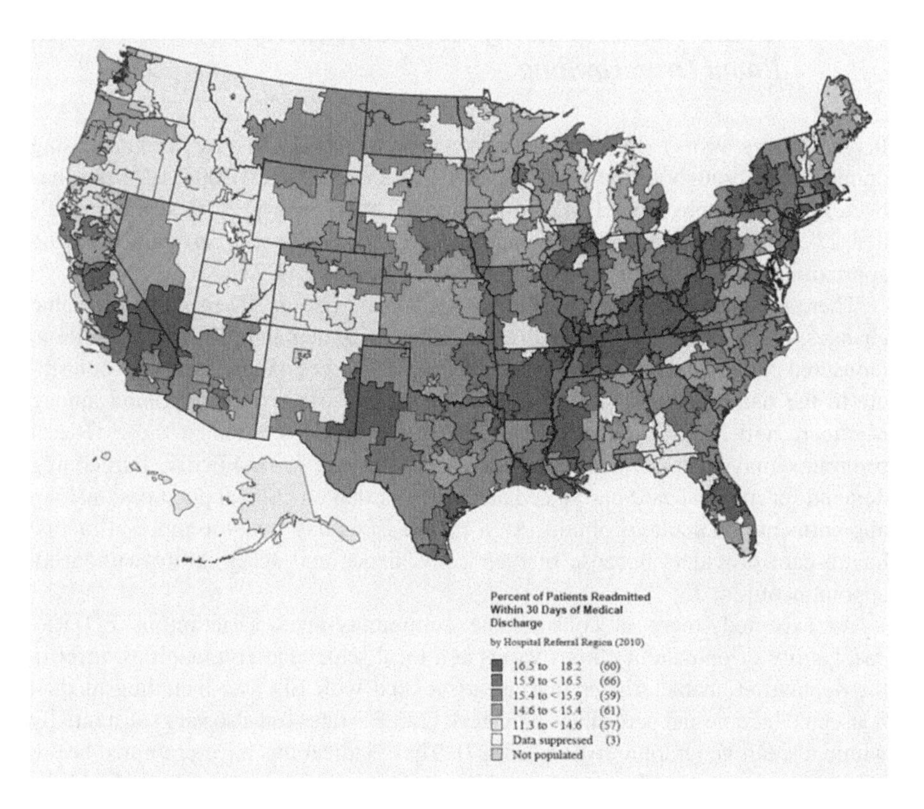

Fig. 10.5 Percent of patients readmitted to a hospital within 30 days of a medical discharge, by hospital referral regions, in 2010 (Source: Goodman et al. [18]. After Hospitalization: A Dartmouth Atlas Report on Readmissions Among Medicare Beneficiaries. The Dartmouth Institute for Health Policy and Clinical Practice)

11.4 % in Ogden, Utah, to 18.1 % in the Bronx, N.Y. Readmission rates were also high in the Detroit (17.8 %) and Chicago (17.7 %) HRRs. Although the study authors emphasized that the readmission rate is an indirect measure of the effectiveness of care coordination, they also recognized that the local patterns of hospital utilization is a poorly understood influence on readmission rates – i.e., health systems that have higher underlying admission rates (suggesting they are more likely to rely on the hospital as a site of care in general) tend to have higher readmission rates. Furthermore, after adjusting for age, gender, race, and the chronic illness mix, the study results suggest that regional patterns of hospital care have a strong effect on hospital readmission rates [18].

Finally, the study concludes that, although some of the variation detected is due to differences in patient populations and local care patterns that keep patients out of the hospital, the regional variability detected does represent significant opportunities for improving care that can lead to fewer hospital days and better outcomes in each HHR.

10.3.4 Leveraging Knowledge of Geographic Variations to Alleviate Community-Level Emergency Room Overcrowding

In recent years, visits to hospital emergency departments (EDs) have been increasing dramatically throughout the United States. Between 1997 and 2007, ED visits has risen 23 %, to approximately 125 million visits every year [19]. Research reveals that 27 % of ED visits could be handled at a retail or urgent care clinic, saving approximately $4.4 billion in healthcare costs annually [20, 21].

There are a number of plausible reasons for the rise in ED use. Demographic changes, such as the aging population, have increased demand for EDs. In addition, uninsured patients, who lack access to alternative sources of care, may also contribute to the rising demand [22]. Furthermore, as ED use is most common among Medicare and Medicaid participants, the growing enrollment in these federal programs may also contribute to increased ED use [23]. Finally, burgeoning demand for medical care has placed an excess burden on clinical practices, making appointments difficult to obtain. As a result, EDs may become more attractive health care providers because of their convenience and accessibility without an appointment [24, 25].

As expected, there is considerable community-level variation in ED use. Availability of outpatient clinics varies at a local scale, and communities differ in the population characteristics that are associated with ED use, including median household income and percentage uninsured [25]. ED rates can also vary substantially within a small geographic region [26, 27]. These indications have prompted health officials and policy makers to seek targeted interventions to identify and address community level disparities in ED use [28].

Knowledge of these geographic variations can contribute to the implementation of community-based initiatives to alleviate ED overcrowding. Identifying areas with high ED use can enable the establishment of community health centers and local urgent care clinics that can provide alternative outlets for primary medical, dental and behavioral healthcare [29]. In addition, mobile health clinics can be deployed (with targeted driving directions) in underserved communities to improve access to basic medical services. Moreover, community outreach teams could be deployed, with geo-enabled patient care plans, in high risk neighborhoods to promote health education, to assist in chronic care management and to provide information about local health resources [30, 31]. Finally, communities can establish more effective modes of transportation to and from local clinics, such as evening and weekend bus, van and car pool services. As many residents may not have access to health services through traditional channels, geographically targeted and enabled community level efforts – such as those described – are essential to alleviating ED overcrowding and reducing ED-related health care costs.

10.3.5 Electronic Medical Records Incorporate Place Histories to Improve Patient Care

In primary care, a clinician's recommendations about healthy lifestyles – such as more physical activity, better diets, avoidance of potential toxins or pollutants – are typically followed in the context of the area in which the patients live. If a care provider suggests that patients increase their intake of fruits and vegetables, but their proximity to a store selling fresh food is low, the chance of compliance will be low. If the medical advice is to walk or bike more for exercise, but their neighborhood has no sidewalks or crosswalk markings, inadequate traffic control, poor lighting, or high crime rates, it is likely that the advice will not be followed.

Consideration of a patient's home and work environment in the delivery of clinical recommendations about specific health promotion and disease prevention goals improves the overall effectiveness of a physician's counseling efforts [32]. Primary care physicians are trained to consider the context of the individual and family, in disease diagnosis, prognosis, and treatment. Today, residency training programs are beginning to train primary care physicians to consider the context of where a person lives as it relates to their health [33]. As healthy lifestyles and chronic disease prevention become more important topics in national health care systems, the opportunities for integrating knowledge of individuals and their families' environments into health assessments, decision making, and treatment become self-evident.

Geo-enabled medical intelligence – such as GIS technology, spatial analysis, data about land use and location of clinical services, and information from our patients about their perceived environment – can be used to better understand the context in which patients live. Place *is* a vital sign – similar to a blood pressure, pulse, or pain score – in providing optimal patient care. Obtaining and using information about place can be part of the qualitative assessment of a patient's home, school, and work environments. As the adoption of electronic health records becomes more widespread, quantitative assessments of a patient's environment (along with the traditional vital signs) will become possible. With the integration of health information and GIS in EMRs, it is possible to integrate data pertaining to location, nearest places for exercise or healthy food, social services, measures of walkability, and other data important in promoting healthy lifestyles [34, 35]. Recommendations can be tailored to each individual patient, taking into account their own habitat and, more importantly, their perceptions of that habitat. A physician can alert a patient to opportunities for healthy lifestyle habits of which they might not have been aware. The discussion of wellness and treatment of illness now becomes personalized, with an enhanced chance of success.

Finally, these tools are not difficult to access. Most practicing physicians have access to Internet resources, including Google Earth and Google Maps. These programs, and others, are GIS, and provide the capability to view multiple data sources in relation to each other across space. Physicians can use these online tools to start understanding the neighborhoods of their patients and the closest available resources that will help shape treatment and decision-making.

10.4 Concluding Remarks

Today, patient care is often delivered by a mix of health care providers, such as primary care physicians, physician assistants, specialists, pharmacists, laboratory clinicians, and insurance companies. Places for service delivery include not only physician offices and local hospitals, but also mini-clinics, employer clinics, and the new patient-centered medical home. From a consumer perspective, patients are now offered coordinated delivery across multiple caregivers, early diagnosis, prompt and reliable delivery, flexible access to care, and price stability. As silos of specialized treatment are known to be insufficient to deliver the desired patient outcomes, economic pressures are forcing many physicians to consolidate into larger group practices, and components of the system are leaning towards integrated health systems [36].

Most people would agree that patient engagement as a critical enabler for delivering a high quality of health care. Indeed, the greatest benefit of a modern-day HIE is enhanced patient engagement – the priceless value of a patient holding his/her primary care record as he/she enters the emergency department for urgent clinical care. A well-designed HIE enables patients to become more educated consumers, who have the power to impact their disease management, diagnosis accuracy, and medication adherence.

Enhancing patient engagement also involves the patients' having suitable geographic access to health care services and delivery. Studies have shown that increasing the distance to health care services results in a reduced utilization of the health care system, and increased (geographically related) inequities in health status [37, 38]. Although distance is a fundamental component of access to care, it is only one factor that might lead to the realization of health services. Other patient-level characteristics that affect whether and where a patient seeks care include age, gender, ethnicity, socioeconomic status, beliefs about health, and the actual need for care [39]. Furthermore, geography and social networks (such as relationships with family, friends, and co-workers) also play a vital role infectious disease transmission [40].

It is evident that the effective control of infectious diseases is dependent on focused efforts to decrease endemic infections in at-risk and susceptible populations. Inherent in these efforts are well-established relationships between the clinical, public health, and research communities, and between medical geographers and the medical community. Collaborative partnerships between medically geographers and public health officials need to be established locally; however, they also need to extend across state, national, and international levels. Indeed, this type of multi-disciplinary and multi-scale approach can result in increased health equity through the design and development of communities that are health-promoting, have more direct access to clinical and preventive services, and refocus the health care system on improving population health [41].

This focus on clinical and community resources is evident in the four approaches that the Centers for Disease Control and Prevention (CDC) has identified as being effective in reducing the burden of chronic disease: (1) epidemiology and surveillance systems that monitor health trends; (2) environmental approaches that promote health and support healthy behaviors; (3) health system interventions to improve the effective use of clinical and other preventive services; and (4) community resources linked to clinical services that sustain improved management of chronic conditions [42].

In the future, through the use of geospatially enabled health surveillance systems and the ability to link community resources and clinical resources to geographical areas with the greatest need, GIS technologies will be able to contribute to an engaged patient care and an improved quality of life. In addition, clinically based community resources will be well-poised to adopt smart HIE technologies that offer patient-centric health services, such as social networking, medical education, and health monitoring, that are tailored to their specific place history and lifestyles.

In conclusion, integration of GIS technologies with modern-day HIEs promises to enhance the quality of experience of patients who are engaged in and accountable for their own health care. Not only does GIS technology provide valuable geospatial health intelligence about environmental risk and exposures, it also brings important insights on the problems of health care inequities and access to care. The advances in information technology – such as mobile personal devices and cost-effective processing and storage solutions – increase the potential of governments, health care systems, and providers to create a community of engaged patients who are able to take a more active role in their own personal health care.

References

1. Stolberg SG, Pear R (2010) Obama signs health care overhaul bill, with a flourish. http://www.nytimes.com/2010/03/24/health/policy/24health.html. Accessed 4 July 2014
2. U.S. Department of Health and Human Services (2011) 2011 Report to congress: national strategy for quality improvement in health care. http://www.ahrq.gov/workingforquality/nqs/nqs2011annlrpt.htm. Accessed 4 July 2014

3. National Prevention Council (2011) National prevention strategy: America's plan for better health and wellness. http://www.surgeongeneral.gov/initiatives/prevention/strategy/. Accessed 4 July 2014
4. U.S. Preventive Services Task Force (2014) http://www.uspreventiveservicestaskforce.org/. Accessed 4 July 2014
5. Advisory Committee on Immunization Practices (2014) http://www.cdc.gov/vaccines/acip/. Accessed 4 July 2014
6. Institute of Medicine (2011) Clinical preventive services for women: closing the gaps http://iom.edu/Reports/2011/Clinical-Preventive-Services-for-Women-Closing-the-Gaps.aspx. Accessed 4 July 2014
7. The Patient Protection and Affordable Care Act of 2010, Section 4002. 42 U.S.C. § 300u-11(a)
8. Centers for Disease Control and Prevention (2014) Meaningful use of electronic records. http://www.cdc.gov/ehrmeaningfuluse/. Accessed 4 July 2014
9. Aylward B, Tangermann R (2011) The global polio eradication initiative: lessons learned and prospects for success. Vaccine 29:D80–D85. doi:10.1016/j.vaccine.2011.10.005
10. Grassly NC (2013) The final stages of the global eradication of poliomyelitis. Philos Trans R Soc B 368:20120140. http://dx.doi.org/10.1098/rstb.2012.0140
11. Centers for Disease Control and Prevention (2014) Polio eradication, microplanning and GIS. http://blogs.cdc.gov/global/2014/06/27/2986/. Accessed 2 Aug 2014
12. Bill and Melinda Gates Foundation (2014) Polio: strategy overview. http://www.gatesfoundation.org/What-We-Do/Global-Development/Polio/. Accessed 2 Aug 2014
13. Carr-Hill R, Hardman G, Martin S et al (1994) A formula for distributing NHS revenues based on small area use of hospital beds. Occasional paper, Centre for Health Economics, University of York
14. Carr-Hill R, Sheldon TA, Smith P et al (1994) Allocating resources to health authorities: development of methods for small area analysis of use of inpatient services. Br Med J 309:1046–1049
15. Smith P, Sheldon T, Martin S (1996) An index of need for psychiatric services based on in-patient utilisations. Br J Psychiatry 169(3):308–316
16. NHS England Strategic Finance (2014) Technical guide to the formulae for 2014–15 and 2015–16 revenue allocations to Clinical Commissioning Groups and Area Teams. NHS England. Available at: http://www.england.nhs.uk/wp-content/uploads/2014/03/tech-guide-rev-allocs.pdf. Accessed 3 Aug 2014
17. Jencks SF, Williams MV, Coleman EA (2009) Rehospitalizations among patients in the medicare fee-for-service program. N Engl J Med 360(14):1418–1428
18. Goodman DC, Fisher ES, Chang C-H (2013) After hospitalization: a Dartmouth Atlas report on readmissions among medicare beneficiaries. The Dartmouth Institute for Health Policy and Clinical Practice, Dartmouth
19. Owens PL, Mutter R (2010) Emergency department visits for adults in community hospitals, 2008. Statistical brief 100. Agency for Healthcare Research and Quality, Rockville
20. Weinick RM, Burns RM, Mehrotra A (2010) Many emergency department visits could be managed at urgent care centers and retail clinics. Health Aff 29:1630–1636
21. Jayaprakash N, O'Sullivan R, Bey T, Ahmed SS, Lotfipour S (2009) Crowding and delivery of healthcare in emergency departments: the European perspective. West J Emerg Med 10:233–239
22. Weber EJ, Showstack JA, Hunt KA, Colby DC, Grimes B, Bacchetti P, Callham ML (2008) Are the uninsured responsible for the increase in emergency department visits in the United States? Ann Emerg Med 52:108–115
23. McCaig LF, Burt CW (2005) National hospital ambulatory medical care survey: 2003 emergency department summary. Advance data from vital and health statistics, no. 358. National Center for Health Statistics, Hyattsville
24. Guttman N, Zimmerman DR, Nelson MS (2003) The many faces of access: reasons for medically non-urgent emergency department visits. J Health Polit Policy Law 28:1089–1120

25. Cunningham PJ (2006) What accounts for differences in the use of hospital emergency departments across U.S. communities? Health Aff 25:324–336
26. Dulin MF, Ludden TM, Tapp H, Smith HA, Urquieta de Hernandez B, Blackwell J, Furuseth OJ (2009) Geographic information systems (GIS) demonstrating primary care needs for a transitioning Hispanic community. J Am Board Fam Med 23:9–20
27. Everage NJ, Pearlman DN, Sutton N, Goldman D (2010) Asthma hospitalization and emergency department visit rates: Rhode Island's progress in meeting healthy people 2010 goals. Health by numbers. Rhode Island Department of Health, Providence
28. Neelon B, Ghosh P, Loebs PG (2013) A spatial Poisson hurdle model for exploring geographic variation in emergency department visits. J R Stat Soc A 176(2):389–413
29. Roby DH, Pourat N, Pirritano MJ, Vrungos SM, Dajee H, Castillo D, Kominski GF (2010) Impact of patient-centered medical home assignment on emergency room visits among uninsured patients in a county health system. Med Care Res Rev 67:412–430
30. Niska R, Bhuiya F, Xu J (2010) National hospital ambulatory medical care survey: 2007 emergency department summary. Report 26. National Center for Health Statistics, Washington, DC
31. Pillow MT, Doctor S, Brown S, Carter K, Mulliken R (2013) An emergency department-initiated, web-based, multidisciplinary approach to decreasing emergency department visits by the top frequent visitors using patient care plans. J Emerg Med 44(4):853–60. http://dx.doi.org/10.1016/j.jemermed.2012.08.020
32. Berke EM (2010) Geographic information systems (GIS): recognizing the importance of place in primary care research and practice. J Am Board Fam Med 23(1):9–12. doi:10.3122/jabfm.2010.01.090119
33. Hayashi AS, Bazemore A, McIntyre J (2011) Transforming community health and primary care education using clinical and administrative data and geographic information systems. J Map Geogr Libr 7(1):61–70. doi:10.1080/15420353.2011.534690
34. Berke EM, Koepsell T, Moudon AV, Hoskins RE, Larson EB (2007) Association of the built environment with physical activity and obesity in older persons. Am J Public Health 97:486–492
35. Moudon AV, Lee C, Cheadle A et al (2007) Attributes of environments supporting walking. Am J Health Promot 21:448–459
36. Ciriello JN, Kulatilaka N (2010) Smart health community: the hidden value of health information exchange. Am J Manage Care 16:SP31–SP36
37. Hiscock R, Pearce J, Blakely T, Witten K (2008) Is neighborhood access to health care provision associated with individual level utilization and satisfaction? Health Serv Res 43(6):2183–2200
38. Korda RJ, Butler JRG, Clements MS, Kunitz SJ (2007) Differential impacts of health care in Australia: trend analysis of socioeconomic inequalities in avoidable mortality. Int J Epidemiol 36(1):157–165
39. Gatrell AC (2002) Geographies of health: an introduction. Wiley-Blackwell, London
40. Emch M, Root ED, Giebultowicz S et al (2012) Integration of spatial and social network analysis in disease transmission studies. Ann Assoc Am Geogr 102(5):1004–1015. doi:10.1080/00045608.2012.671129
41. Khabbaz RF, Moseley RR, Steiner RJ, Levitt AM, Bell BP (2014) Challenges of infectious diseases in the USA. Lancet 384(9937):53–65. doi:10.1016/S0140-6736(14)60890-4
42. Bauer UE, Briss PA, Goodman RA, Bowman BA (2014) Prevention of chronic disease in the 21st century: elimination of the leading preventable causes of premature death and disability in the USA. Lancet 384(9937):45–52. doi:10.1016/S0140-6736(14)60648-6

Index

A.J. Blatt, *Health, Science, and Place*, Geotechnologies and the Environment 12,
DOI 10.1007/978-3-319-12003-4

CPSIA information can be obtained at www.ICGtesting.com
Printed in the USA
LVOW01*1700241114

415380LV00001B/2/P

9 783319 120027